Book of Strains
Second Edition
A Professor Grow Guide

The 65 Cannabis Strains Most Commonly Found at Dispensaries

The information contained in this book is not intended to diagnose, treat, cure or prevent any disease. Be sure to check with your health care practitioner before using medical marijuana.

Medical Marijuana use is NOT approved in all states. The information contained in this book is for patients and caregivers who are operating within their state and local laws. We do not advocate unlawful use of cannabis. Marijuana remains illegal under Federal Law. Use of marijuana, even for medical purposes carries a potential risk of arrest.

Be sure to review your local laws and obtain necessary permits before using medical marijuana.

Book of Strains
Second Edition
A Professor Grow Guide

The 65 Cannabis Strains Most Commonly Found at Dispensaries

Justin Griswell and Victoria Young

Book of Strains, 2nd Edition
A Professor Grow Guide
Copyright © 2013 by Justin Griswell and Victoria Young
ISBN-13: 978-1-940548-00-5
ISBN-10: 1940548004

All rights reserved. No part of this book may be reproduced in any form without written permission.

For information, contact:

Professor Grow, LLC
P. O. Box 366
Firestone, CO 80520-0366
professorgrow@gmail.com

Published by Professor Grow, LLC
Printed in the United States of America
Second Edition: August 2013

Book Design by Victoria Young
Cover design by Victoria Young

This book is also available on Kindle.

Visit us online at: http://professorgrow.com

eMail us at: professorgrow@gmail.com.

Watch for our forthcoming books and classes on Organic Gardening, Bonsai and Growing Medical Marijuana.

For Brother Duke and Pat....

*And everyone
who is struggling with pain,
illness and depression.*

Foreword

Professor Grow's Book of Strains grew organically out of the lives of the two authors. It was a natural progression of circumstances. This is the second edition of the book, in which we have added more strains.

Victoria Young was diagnosed with an inherited neuromuscular disease in 1990. The symptoms include muscle wasting and severe nerve pain. Traditional medicine offered drugs that made her uncomfortable at best and very sick at worst. Drugs that were strong enough to help her sleep, left her drowsy, lethargic and unable to be productive. Other drugs caused extreme nausea to the point that she could not do anything at all.

Victoria resolved to find more natural ways to deal with pain. She altered her diet to eliminate or reduce allergens such as wheat, soy and dairy. It helped a little, but the pain was still there. She still had shooting pains that came without warning. Meditation helped a little, too.

Over time and with practice she was able to "ignore" the pain during the day when she was engaged in projects or something that could occupy her attention. The pain was still there, but she could put it on the "back burner" so-to-speak. Having her mind actively engaged in something took her mind off the pain.

But night was a different story. In order to get to sleep, she needed to "shut down" her mind. When she began to shut down her mind for sleep, all that was left was the pain.

Eventually, after several people suggested it, she got a doctor's recommendation to use marijuana and registered with the state, getting a card that enabled her to visit dispensaries and purchase cannabis. She was quite surprised at how well it helped her sleep. And she had no "hangover" the next day. It was welcome relief that she used only at night. There were some strains that caused a headache so she had to spend time trying strains to see which ones she tolerated well. She discovered it was Indica strains causing headaches.

Her lungs could not tolerate smoke so she made brownies (wheat-free). It requires significantly more cannabis to make edibles than to smoke it for effect. And prices in the dispensaries were averaging $400 per ounce. Her best alternative became to grow her own medicine. Colorado, the state in which she resides, allows her to grow up to six plants at a time. A maximum of three can be flowering at one time.

Justin Griswell is Victoria's son. He is her designated caregiver registered with the state of Colorado. He had been growing and selling Bonsai trees for more than a dozen years. He also grows organic gardens and is very active with cultivating trees flowers and food. His knowledge of plants, growing nutrients and light is extensive.

Justin had studied many books and articles on growing cannabis and was able to put his considerable knowledge to work and assist Victoria in growing her own medicine.

During many months of experimentation with three plants at a time, they tried growing in soil and hydroponics. They eventually built their own grow lights, built their own hydroponics system and built their own screens to reflect and control light.

Their grow lights are NOT the typical HID lights that burn very hot and require great care and ventilation. Their CFL grow lights provide plenty of light (800 watts equivalent) for 3 plants and produce a nice yield. They do not require ventilation or heat mitigation. They designed and built a 6500k version for the vegetative state and a 2700k version for flowering.

Their homemade hydroponics system accommodates three plants nicely and produces a decent yield. They can harvest three plants every two to three months and stay within the state guidelines. Justin

Foreword

became inspired at the idea of helping patients (many of whom are disabled and on limited incomes) grow their own medicine safely and inexpensively (compared to dispensary prices).

During this process Justin and Victoria met other MMj patients who had dozens of questions about growing. Justin answered all of their questions with good, solid, easy-to-understand information. As a result, one of those people suggested that Justin call himself "Professor Grow" and teach this to others.

Victoria is in the domain name business so she quickly registered professorgrow.com and they began the process of developing Professor Grow LLC. Their mission is to teach growing classes, write books and create inexpensive "grow kits" that will enable patients and caregivers grow cannabis as medicine safely, inexpensively and easily within the laws of their individual states.

This book is the first in a series that we hope will help people figure out how to choose their strains and grow them as medicine where it is allowable by law.

Preface

We were already working on another book about growing and working on outlines and PowerPoint presentations for growing classes when the idea for this book intervened. This second addition adds 15 strains.

I (Victoria) was still seeking ideal strains to treat my nerve pain so I could sleep at night. I wanted to identify up to 5 strains that worked well for me without side effects. I wanted to have other strains to switch to if my body built up a tolerance to one strain and it stopped working well.

The dispensaries kept guiding me to indica dominant strains and I kept getting headaches from most of them. Justin and I are still pondering whether it was the actual strains themselves, or perhaps some chemicals used in their growing. (Another reason to grow my own – we grow organically, whether it is food or medicine.)

I was trying to do some more research on strains without having to buy and try a sample of every strain available. My research included reading the descriptions dispensaries posted on websites. I found the descriptions to be pure marketing speak. No information in the description was related to what I wanted to know. The description would tell me how it looked, how it tasted and how much of a buzz I would get.

I wanted to know if it was supposed to be good for nerve pain, if it would help me sleep and if I would have lasting effects that would impair me in the morning. If I encountered a day in which I needed to use MMj in the daytime, I wanted a strain that would not give me "couchlock." And I needed to make sure any daytime strain I occasionally used would not interfere with my ability to focus and be productive. I did not want to be sitting at the computer with no ability to concentrate on what I was doing. I know from my own experience that some strains enable focus and some strains impair concentration.

Preface

Other research on the internet produced unsatisfactory results, as well. The myriad of "stoner forums" carried reviews (I use the term loosely) related to how messed up someone could get using it. They used strings of profanity to describe how totally stoned they became. (And they wonder why the mainstream hasn't accepted cannabis.) There were a few posts on some forums that gave some relevant information, but they were few and far between.

I told Justin that I wish we had the information that was valuable to me gathered up in one place. As we discussed it, we decided that it would be helpful for many Medical Marijuana users. Thus, Book of Strains was born.

There are places on the internet that say there are over 1000 strains of cannabis. We decided to concentrate on the 65 strains that we most commonly see in dispensaries. Since we are in Colorado, we polled Colorado dispensaries. We also looked at the strain menus of many dispensaries in California and Washington. Although new strains may develop and popularity of some strains may wane, we believe these 65 strains represent the strains (and clones) that are most commonly available to MMj patients who visit dispensaries.

We plan to publish revised editions of this book when it is necessary to remain accurate and relevant.

DISCLAIMER: The information contained in this book is not intended to diagnose, treat, cure, or prevent any disease. Be sure to check with your health care practitioner before using medical marijuana.

Medical Marijuana use is NOT approved in all states. The information contained in this book is for patients and caregivers who are operating within their state and local laws. We do not advocate unlawful use of cannabis. Be sure to review your local laws, and obtain necessary permits before using medical marijuana. Use of marijuana, even for medicinal purposes, remains illegal under Federal law.

Introduction

We wrote this book to help MMj patients make more informed choices when going to a dispensary or choosing what to grow. We wanted to gather information into one easy reference for Medical Marijuana Patients and Caregivers. The information contained herein is on the 65 Cannabis Strains most commonly found in dispensaries. We know these are the 65 strains to which most patients will have access in usable form(s) and as clones to grow.

When we started, we were open to covering as many strains as necessary in order to cover the most commonly available cannabis strains. Our research has produced 65 strains that are the most commonly available in dispensaries, in usable form and/or clones.

We have organized the information with a two-page spread for each strain that contains relevant information for patients and caregivers. We have also provided indices that sort the information by the diseases for which people seek relief using cannabis.

And there is an index listing Ailments and Medical Marijuana strains that have historically been used for those ailments. Another index lists testing labs where you can get your strain(s) tested for potency, cannabinoid content and contaminants.

We recommend that you take some time to figure out what strains work best for you before you begin to grow. Visit a reputable dispensary or licensed caregiver or licensed grower to try small amounts of strains (buy only a gram of a new strain) to find out what will work for you. It is unnecessary to lay out a significant amount of money to find out a strain doesn't work for your needs.

There are some <u>notable limitations</u> in this book:

- We cover only 65 strains out of over 1000 that are purported exist.

Introduction

In our research we found that only 65 strains are very common in dispensaries. Individual dispensaries may carry a strain not found in this book, but it is not widely found at many dispensaries. We made the conscious decision to limit the scope of this book to the strains that one would most often find at a dispensary.

- None of the information in this book is intended to replace the advice of your licensed medical practitioner nor is it intended to diagnose, treat, cure or prevent any disease. Check with your licensed medical practitioner before using medical marijuana.

- We cannot guarantee the THC content of any strain. Our sources for info on THC content and genetics are seed banks' published strain info on the internet and information acquired from dispensaries, and strain testing results from labs that publish test results. We believe the THC info is reliable but cannot guarantee its accuracy. THC Content in one strain can vary dramatically due to differences in growing environments and growing methods. The only accurate way to ascertain true THC content is by having a sample tested by a lab. You can also ask your dispensary if they have had their strain(s) tested by a lab, and what the results were.

- There are multiple strains with the same name or multiple versions of some strains. We have seen AK-47 listed as 80% sativa at one dispensary and as 65% sativa at another. We have also seen Jilly Bean listed as 80% Indica in one dispensary and as 80% Sativa in another. In case of discrepancies we, looked at multiple sources and used the most prevalent response. We believe our information is accurate, but cannot guarantee the accuracy of Sativa/Indica content.

- We cannot guarantee that the strain of usable medicine clones or seeds you acquire are really the strain they are purported to be. There could be errors in labeling etc. We believe the majority of dispensaries and other licensed MMj outlets are carefully tracking their product (required in Colorado) so you should be able to rely on their statements but we cannot be responsible for any errors (they are human, after all) that your source may make.

- You cannot always know if a strain is pure or stable. Especially if you are acquiring your seeds or clones from an individual grower. We have seen cases where a patient acquired a clone of Hash Plant that was in no way similar to a Hash Plant clone acquired elsewhere. One of them could have been a cross or an unstable strain of Hash Plant that caused it to have different traits altogether.

- Medical Marijuana Laws in states including approved ailments and rules on growing are constantly changing. Therefore, we provide (at the back of the book) a list of organizations that keep track of the laws and pending laws in the U.S. states. To find the latest info on your state's marijuana laws, visit the websites' addresses provided on the page titled "***Marijuana Laws***."

- This book is not for recreational cannabis users in states where recreational cannabis is not legal or for anyone who intends unlawful use of cannabis. We are providing this information as a service to medical marijuana patients and caregivers.

About Sativas and Indicas

When you are choosing your strain(s), there are a few generalities that are usually true about Sativas and Indicas. Keep in mind that each person's reaction to a particular strain can be quite different than the "norm," depending on the individual's circumstances.

Growing

Indicas tend to grow shorter (usually under 6 feet), bushier and have a shorter flowering time. Their leaves are more broad and full and deep green, sometimes with a purple tinge. Indicas tend to have more bud sites and usually produce smaller buds than Sativas. Indicas are usually more suited to indoor growing because of their shorter stature.

Sativas tend to grow tall (most are 8 to 12 feet, but some can grow to 25 feet), they have a longer flowering time and usually produce long buds. Most Sativas are more suited to outdoor growing because of their sheer size. However, some breeders have developed Sativa dominant strains that grow smaller and finish quicker more like an Indica.

Check out our books *__Growing Indica__*, *__Growing Sativa__* and *__Growing Auto-Flowers__* for more info on growing various strains.

Body Effects

Indicas have more full body effects and are touted for pain, insomnia, loss of appetite, etc. They are more often consumed at night, when getting ready for sleep. Though this is generally true, I (Victoria) have found that Indicas give me headaches and neck and shoulder pain. I have met a few other women for whom this is true. I don't know if this is a common problem, or just a fluke.

Sativas have a more cerebral effect. They can improve mood, give a feeling of optimism and well-being, as well as some pain relief. Sativas are often recommended for depression, anxiety, increasing creativity, and are usually consumed during the day time. As stated before, because of the headache issues that I have with Indicas I use Sativas for pain relief and insomnia.

Table of Contents

Foreword i

Preface iv

Introduction vi

About Sativas and Indicas ix

The Strains 1

Afghan Goo (aka Afghooey) 2

Afghooey Train Wreck 4

AK-47 6

Banana Kush 8

Big Bud 10

Blackberry Kush 12

Blue Dream 14

Blue Moonshine 16

Blueberry 18

Bruce Banner 20

Bubba Kush 22

Bubble Berry 24

Bubble Gum 26

Buddha's Sister 28

Champagne 30

Chem Dog (aka Chem Dawg) 32

Table of Contents

Chernobyl 34

Chocolope 36

Critical Mass 38

Diablo OG 40

Durban Poison 42

Flo 44

G-13 46

G-13 x Haze 48

Girl Scout Cookies 50

God's Gift 52

Grand Daddy Purple 54

Grapefruit 56

Great White Shark 58

Hash Plant 60

Hawaiian 62

Hawaiian Skunk 64

Hawaiian Snow 66

Headband 68

Island Sweet Skunk 70

Jack Flash 72

Jack Herer 74

Jilly Bean 76

Table of Contents

Juicy Fruit 78

LA Confidential 80

Lambsbread 82

Lavender Kush 84

Lemon Skunk 86

Mango Haze 88

Master Kush 90

Maui Wowie 92

Northern Lights 94

NYC Diesel 96

OG Kush 98

Purple Haze 100

Shiva Shanti 102

Silver Haze 104

Skunk 106

Skywalker 108

Sour Diesel 110

Space Queen (AKA Space Jill) 112

Strawberry Cough 114

Strawberry Diesel 116

Super Lemon Haze 118

Super Silver Haze 120

Table of Contents

Sweet Tooth 122

Train Wreck 124

Vanilla Kush 126

White Rhino 128

White Widow 130

Marijuana Laws 133

Ailment Index 135

Ailments and Strains 137

Cannabis Analysis Labs 155

The Strains

Afghan Goo (aka Afghooey)

Lineage: Maui Haze x Afghani #1

Genetics: 80% (I)   20% (S)

THC Content: 16% — 20%

Past Medicinal Uses*:

AIDS
Alcoholism
Anorexia
Anxiety
Appetite
Arthritis
Cancer
Depression
Epilepsy
Gastrointestinal Issues
Glaucoma
Insomnia
Migraines
MS
Muscle Spasms
Nausea
Pain
PMS
PTSD

Growing: Easy

Flowering Time: 60 - 65 Days

Description: This Indica dominant strain doesn't get very tall. Its short stature makes it a good candidate for growing in small places.

It is a good plant to top two or three times during the vegetative stage to produce more budding sites. You can find out more about topping your plants by visiting:

http://professorgrow.com/2010/09/21/mmj-corner-topping-and-cropping-to-increase-your-yield/

Afghan Goo could be a good strain for a beginning grower, because it is an easy plant to manage in small spaces.

*The *Past Medicinal Uses* information in this book is NOT intended to replace the advice of your licensed medical practitioner nor is it intended to diagnose, treat, cure or prevent any disease. It is merely a listing of some historical uses for this particular strain. Contact your licensed medical professional for advice on using various strains for your ailments.

Afghooey Train Wreck

Lineage: Afghani x Columbian x Mexican

Genetics: 50% (S)   50% (I)

THC Content: 15% — 20%

Past Medicinal Uses*:

Appetite Stimulation

Cramps

Depression

Fibromyalgia

Joint Pain

Migraines

Muscle Pain

Nausea

Pain Relief

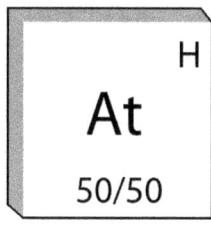

Growing: Easy

Flowering: 42 - 67 Days

Description: This strain is a 50/50 mix. We would suggest topping the plant three to four times during the vegetative stage. (For information on topping plants, visit http://professorgrow.com.) This is because of the possible height the plant may get.

Afghooey Train Wreck takes after its Sativa genetics in the height department. Topping helps keep the height under control and bush the plant out a bit. This can be a good candidate for smaller indoor grow spaces, since this strain is a 50-50 mix. The flowering period will vary if you are growing from seeds. If you are growing from clones then you will have a more uniform growing and flowering.

We recommend growing this strain from clones.

*The *Past Medicinal Uses* information in this book is NOT intended to replace the advice of your licensed medical practitioner nor is it intended to diagnose, treat, cure or prevent any disease. It is merely a listing of some historical uses for this particular strain. Contact your licensed medical professional for advice on using various strains for your ailments.

AK-47

Lineage: [Columbian x Mexican] x [Thai x Afghani]

Genetics: 65% (S)   35% (I)

THC Content: 18% — 20%

Awards: 1999 Sativa Cup 2nd Place

Past Medicinal Uses*:

Alzheimer's Disease
Bipolar Disorder
Chronic Pain
Depression
Headache
Nausea
Poor Appetite
Vomiting

Growing: Moderate

Flowering: 60 - 75 Days

Description: AK-47 is Sativa dominant. It tends to grow tall. We would suggest topping the plant three to four times during the vegetative stage because of limited side branching.

You can find out more about topping your plants by visiting:

http://professorgrow.com/2010/09/21/mmj-corner-topping-and-cropping-to-increase-your-yield/

This plant will need a little more room to grow. When putting in to your grow room keep this in mind. If you have a small grow room, we suggest you only grow two plants because of height, and lengthy flowering time.

*The *Past Medicinal Uses* information in this book is NOT intended to replace the advice of your licensed medical practitioner nor is it intended to diagnose, treat, cure or prevent any disease. It is merely a listing of some historical uses for this particular strain. Contact your licensed medical professional for advice on using various strains for your ailments.

Blackberry Kush

Lineage: Afghani x Blackberry

Genetics: 75% (I)   25% (S)

THC Content: 15% — 24%

Awards: 2010 Spannabis Resin Cup — 2nd Place

Past Medicinal Uses*:

Anxiety

Appetite

Depression

Insomnia

Muscle Spasms

Nausea

Stress

The 65 Most Common Strains

Growing: Easy

Flowering: 60 - 74 Days

Description: Blackberry Kush is a good Indica dominant strain to grow in a small room, but expect it to produce smaller buds compared to other Kush strains. This plant won't get very tall during vegetative stage. During flowering stage you will see it begin to produce its purple buds. We would suggest topping the plant one or two times during its vegetative stage to create more budding sites. Three plants will do well in small grow room. You can find out more about topping your plants by visiting:

http://professorgrow.com/2010/09/21/mmj-corner-topping-and-cropping-to-increase-your-yield/

If you want more side branching, you can top it more if you have a large enough grow area.

*The _Past Medicinal Uses_ information in this book is NOT intended to replace the advice of your licensed medical practitioner nor is it intended to diagnose, treat, cure or prevent any disease. It is merely a listing of some historical uses for this particular strain. Contact your licensed medical professional for advice on using various strains for your ailments.

Banana Kush

Lineage: OG Kush x Banana

Genetics: 60% (I) 40% (S)

THC Content: 18% — 20%

Past Medicinal Uses*:

Eating Disorders
Fibromyalgia
Glaucoma
Insomnia
Joint Pain
Muscle Spasms
Nausea
Pain Relief
Phantom Limb Pain

The 65 Most Common Strains

Growing: Moderate

Flowering: 55 - 60 Days

Description: Banana Kush is Indica dominant. It won't get very tall. We suggest topping it three to four times during the vegetative stage to produce more budding sites. Banana Kush is a good candidate for small grow rooms, because of its shorter stature.

You can find out more about topping your plants by visiting:

http://professorgrow.com/2010/09/21/mmj-corner-topping-and-cropping-to-increase-your-yield/

Just keep a good eye on the plant during the flowering period. The flowers tend to grow very close to the main stock.

*The *Past Medicinal Uses* information in this book is NOT intended to replace the advice of your licensed medical practitioner nor is it intended to diagnose, treat, cure or prevent any disease. It is merely a listing of some historical uses for this particular strain. Contact your licensed medical professional for advice on using various strains for your ailments.

Big Bud

Lineage: Skunk #1 x Afghani

Genetics: 75% (I)   25% (S)

THC Content: 8% — 15%

Past Medicinal Uses*:

Anxiety
Appetite
Depression
Insomnia
Migraines
MS
Nausea
Stomach Pains
Stress
Tourette's Syndrome

The 65 Most Common Strains

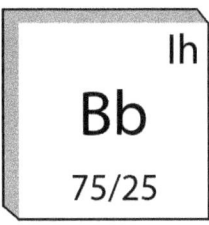

Growing: Easy

Flowering: 60 - 74 Days

Description: Big Bud is a great Indica dominant strain to grow in a small room, and produce good medicine yields. This plant won't get very tall during vegetative stage. During flowering stage you will want to give the plant plenty of room to produce its large buds. We would suggest to only grow two plants in a small grow room with the amount of medicine you produce. This plant has great natural side branching, so we would only suggest topping two to three times in a small grow room. You can find out more about topping your plants by visiting:

http://professorgrow.com/2010/09/21/mmj-corner-topping-and-cropping-to-increase-your-yield/

If you want more side branching, you can top it more if you have a larger grow area.

*The *Past Medicinal Uses* information in this book is NOT intended to replace the advice of your licensed medical practitioner nor is it intended to diagnose, treat, cure or prevent any disease. It is merely a listing of some historical uses for this particular strain. Contact your licensed medical professional for advice on using various strains for your ailments.

Blue Dream

Lineage: Blueberry x Haze

Genetics: 80% (S)   20% (I)

THC Content: 15% — 20%

Past Medicinal Uses*:

AIDS
Anorexia
Appetite
Arthritis
Cancer
Depression
Epilepsy
Gastrointestinal Issues
Glaucoma
MS
Muscle Spasms
Nausea
Pain Relief
PTSD

Growing: Moderate

Flowering: 60 - 70 Days

Description: Blue Dream is Sativa dominant. This strain can get very tall. We suggest topping four to five times during the vegetative stage in order to keep the height in a good level for the flowering stage. A good rule of thumb is that the pure Sativa and heavily Sativa dominant strains will double in height during flowering. So, if you put it into flowering at 3 feet tall, it will be about 6 feet tall at harvest time. Frequent topping produces more side branches and keeps height under control. You can find out more about topping your plants by visiting:

http://professorgrow.com/2010/09/21/mmj-corner-topping-and-cropping-to-increase-your-yield/

This is not a good strain for a small grow room, because of height and flowering time. You could grow one Blue Dream plant in a smaller grow room.

*The *Past Medicinal Uses* information in this book is NOT intended to replace the advice of your licensed medical practitioner nor is it intended to diagnose, treat, cure or prevent any disease. It is merely a listing of some historical uses for this particular strain. Contact your licensed medical professional for advice on using various strains for your ailments.

Blue Moonshine

Lineage: Highland Thai x Afghan

Genetics: 90% (I)   10% (S)

THC Content: 15% — 21%

Past Medicinal Uses*:

Anorexia

Anticonvulsant

Anxiety

Appetite Stimulation

Depression

Insomnia

Nausea

Pain Relief

Stress

Growing: Easy

Flowering: 45 - 55 Days

Description: Blue Moonshine is almost a pure Indica. It tends to stay small. We would suggest topping the plant two to three times during the vegetative stage to create more budding sites. You can find out more about topping your plants by visiting:

http://professorgrow.com/2010/09/21/mmj-corner-topping-and-cropping-to-increase-your-yield/

Because of its short stature, Blue Moonshine is ideal for the small grow room. With heavy topping three plants will do well in a small grow room.

*The *Past Medicinal Uses* information in this book is NOT intended to replace the advice of your licensed medical practitioner nor is it intended to diagnose, treat, cure or prevent any disease. It is merely a listing of some historical uses for this particular strain. Contact your licensed medical professional for advice on using various strains for your ailments.

Blueberry

Lineage: Purple Thai x Afghan

Genetics: 80% (I)   20% (S)

THC Content: 15% — 21%

Awards: 2000 Cannabis Cup 1st Place
2000 Indica Cup 1st Place
2001 Cannabis Cup 2nd Place
2001 Indica Cup 3rd Place

Past Medicinal Uses*:

AIDS
Anorexia
Anxiety
Chronic Pain
Depressive Disorder
Diarrhea
Emotional Lability
Gastrointestinal Disorder
Insomnia
Muscle Pain
Muscle Spasm
Nausea
Pain

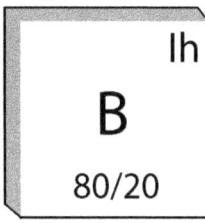

Growing: Easy

Flowering: 45 - 55 Days

Description: Blueberry is an Indica dominant strain that does very well in small grow rooms. The plant tends to stay short. We would suggest topping the plant three to four times during the vegetative stage.

This will help produce more budding sites. You can find out more about topping your plants by visiting:

http://professorgrow.com/2010/09/21/mmj-corner-topping-and-cropping-to-increase-your-yield/

Blueberry has good side branching, so if you don't want top it, you will still produce a good amount of medicine.

*The *Past Medicinal Uses* information in this book is NOT intended to replace the advice of your licensed medical practitioner nor is it intended to diagnose, treat, cure or prevent any disease. It is merely a listing of some historical uses for this particular strain. Contact your licensed medical professional for advice on using various strains for your ailments.

Bruce Banner

Lineage: OG Kush x Strawberry Diesel x Ft. Collins Cough

Genetics: 80% (S)   20% (I)

THC Content: 16% -— 23%

Past Medicinal Uses*:

Anxiety

Arthritis

Depression

Glaucoma

Insomnia

Migraines

Nausea

Pain

Growing: Moderate

Flowering: 56 - 70 Days

Description: Bruce Banner is a Colorado-developed Sativa dominant strain. It tends to grow to a medium to tall height.

We would suggest topping the plant three to four times during the Vegetative stage to give it more side branching. You can find out more about topping your plants by visiting:

http://professorgrow.com/2010/09/21/mmj-corner-topping-and-cropping-to-increase-your-yield/

This plant will need a little more room to grow. Keep this in mind if you have a small grow room. We suggest you only grow two plants because of height, and its potential to take up space.

*The *Past Medicinal Uses* information in this book is NOT intended to replace the advice of your licensed medical practitioner nor is it intended to diagnose, treat, cure or prevent any disease. It is merely a listing of some historical uses for this particular strain. Contact your licensed medical professional for advice on using various strains for your ailments.

Bubba Kush

Lineage: Bubble Gum x Kush

Genetics: 60% (I)   40% (S)

THC Content: 15% — 18%

Past Medicinal Uses*:

Aches
Anxiety
Headaches
Insomnia
Pain
Soreness
Stiff Muscles

Growing: Easy

Flowering: 60 - 70 Days

Description: Bubba Kush is an Indica dominant strain. It tends to grow to a medium height. We would suggest topping two to three times during the vegetative stage because of low side branching. You can find out more about topping your plants by visiting:

http://professorgrow.com/2010/09/21/mmj-corner-topping-and-cropping-to-increase-your-yield/

This is a good plant for a small grow room. Three plants would do well in a smaller growing area. Just keep in mind that Bubba Kush does have a little longer flowering period than most Indicas. Remember to top it a few times to increase side branching and increase bud sites for a better yield.

*The *Past Medicinal Uses* information in this book is NOT intended to replace the advice of your licensed medical practitioner nor is it intended to diagnose, treat, cure or prevent any disease. It is merely a listing of some historical uses for this particular strain. Contact your licensed medical professional for advice on using various strains for your ailments.

Bubble Berry

Lineage: Bubblegum x Blueberry

Genetics: 80% (S)   20% (I)

THC Content: 15% — 20%

Past Medicinal Uses*:

ADD
Depression
Epilepsy
Insomnia
Loss of Appetite
Migraines
Muscle Tension
Pain Relief
Stomach Ulcers
Stress

The 65 Most Common Strains

Growing: Easy

Flowering: 50 - 60 Days

Description: Bubble Berry is a Sativa dominant strain. It tends to grow to a medium height. The good thing about this strain is the shorter than usual flowering time for a Sativa dominant strain.

This strain can be grown in a small grow room. You probably should grow no more than two plants in a small growing area, due to the size of the Sativa strain.

We suggest that you top these plants three to four times during the vegetative cycle. You can find out more about topping your plants by visiting:

http://professorgrow.com/2010/09/21/mmj-corner-topping-and-cropping-to-increase-your-yield/

*The *Past Medicinal Uses* information in this book is NOT intended to replace the advice of your licensed medical practitioner nor is it intended to diagnose, treat, cure or prevent any disease. It is merely a listing of some historical uses for this particular strain. Contact your licensed medical professional for advice on using various strains for your ailments.

Bubble Gum

Lineage: Big Skunk x NL #5

Genetics: 60% (S)   40% (I)

THC Content: 15% — 20%

Awards: 1995 Cannabis Cup 2nd Place

Past Medicinal Uses*:

ADD
Alzheimer's Disease
Anticonvulsant
Anxiety
Appetite Stimulant
Depression
Glaucoma
Muscle Spasms
Nausea
Stress

Growing: Moderate

Flowering: 56 - 63 Days

Description: Bubble Gum is a Sativa dominant strain. It tends to grow tall. We would suggest topping the plant three to four times during the vegetative stage to keep it to a manageable height. For information on topping plants, visit:

http://professorgrow.com /2010/09/21/mmj-corner-topping-and-cropping-to-increase-your-yield/)

This plant will need a little more room to grow. When putting Bubble Gum into your grow room keep this in mind, especially if you have a small grow room. We suggest you only grow two plants because of the height, and flowering time of this strain.

*The *Past Medicinal Uses* information in this book is NOT intended to replace the advice of your licensed medical practitioner nor is it intended to diagnose, treat, cure or prevent any disease. It is merely a listing of some historical uses for this particular strain. Contact your licensed medical professional for advice on using various strains for your ailments.

Buddha's Sister

Lineage: Reclining Buddha x Afghani Hawaiian

Genetics: 80% (I)   20% (S)

THC Content: 15% — 20%

Awards: 2002 Indica Cup 2nd Place

Past Medicinal Uses*:

ADD
ADHD
ALS
Appetite Stimulation
Body Aches
Depression
Epilepsy
Insomnia
Lou Gehrig's Disease
Nausea

Growing: Moderate

Flowering: 67 - 75 Days

Description: Buddha's Sister is an Indica dominant strain. It tends to grow to a medium height. It also has good natural side branching.

Because of the natural side branching, you don't need to top this strain as much as others. You may still want to top it once or twice during the vegetative stage. Find out more about topping your plants by visiting:

http://professorgrow.com/2010/09/21/mmj-corner-topping-and-cropping-to-increase-your-yield/

Buddha's Sister has a longer flowing time than most plants with this much Indica in them, so keep that in mind if choosing to grow it. This is a good plant for a small grow room. Two to three plants will do well. Keep the side branching in mind when growing in a small grow room.

*The *Past Medicinal Uses* information in this book is NOT intended to replace the advice of your licensed medical practitioner nor is it intended to diagnose, treat, cure or prevent any disease. It is merely a listing of some historical uses for this particular strain. Contact your licensed medical professional for advice on using various strains for your ailments.

Champagne

Lineage: [Hash Plant x Hindu Kush] x Burmese

Genetics: 50% (I) 50% (S)

THC Content: 18% — 20%

Past Medicinal Uses*:

ADD
ADHD
Aids
Anxiety
Autism
Bipolar Disorder
Cancer
Depression
Epilepsy
Fibromyalgia
Gastrointestinal Disorder
Glaucoma
Inflammation
Migraines
Pain Relief
PMDD
PMS
PTSD
Seizures

Growing: Moderate

Flowering: 60 - 75 Days

Description: Champagne is a 50/50 mix of Indica and Sativa. This plant tends to grow to a medium-tall height.

We would suggest topping three to four times during the Vegetative stage. This will help keep the plant at a good height. You can find out more about topping your plants by visiting:

http://professorgrow.com/2010/09/21/mmj-corner-topping-and-cropping-to-increase-your-yield/

This plant will need a more room to grow. Keep this in mind if you have a small grow room. We suggest you only grow two plants unless you have a big grow area.

*The *Past Medicinal Uses* information in this book is NOT intended to replace the advice of your licensed medical practitioner nor is it intended to diagnose, treat, cure or prevent any disease. It is merely a listing of some historical uses for this particular strain. Contact your licensed medical professional for advice on using various strains for your ailments.

Chem Dog (aka Chem Dawg)

Lineage: OG Kush x Sour Diesel

Genetics: 60% (I)   40% (S)

THC Content: 15% — 20%

Past Medicinal Uses*:

Anxiety
Deep Muscle Pain
Glaucoma
Joint Pains
Mood Elevation
Movement Disorders
Nausea
Restless Behaviors

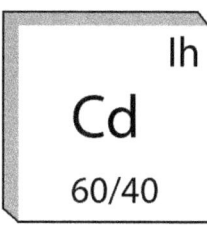

Growing: Easy-Moderate

Flowering: 60 - 70 Days

Description: Chem Dog is an Indica dominant strain. It tends to grow to a taller height than most Indicas.

We would suggest topping the plant three to four times during the vegetative stage. You can find out more about topping your plants by visiting:

http://professorgrow.com/2010/09/21/mmj-corner-topping-and-cropping-to-increase-your-yield/

This plant will need a little more room to grow. When putting Chem Dog into your grow room, keep this in mind, especially if you have a small grow room. We suggest you only grow two plants in a small grow room because of height, and flowering time.

*The *Past Medicinal Uses* information in this book is NOT intended to replace the advice of your licensed medical practitioner nor is it intended to diagnose, treat, cure or prevent any disease. It is merely a listing of some historical uses for this particular strain. Contact your licensed medical professional for advice on using various strains for your ailments.

Chernobyl

Lineage: Trainwreck x Trinity x Jack the Ripper

Genetics: 60% (S)   40% (I)

THC Content: 17% — 20%

Past Medicinal Uses*:

Chronic Pain
Depression
Glaucoma
Insomnia
Muscle Spasms
Muscle Tension
Nausea
Pain Relief
Stress

Growing: Easy-Moderate

Flowering: 60 - 70 Days

Description: Chernobyl is a Sativa dominant strain. This plant likes to grow tall.

We suggest topping the plant three to four times during the Vegetative stage to help control height for the grow area. You can find out more about topping your plants by visiting:

http://professorgrow.com/2010/09/21/mmj-corner-topping-and-cropping-to-increase-your-yield/

This plant will need a more room to grow. Keep this in mind if you have a small grow room. We suggest you only grow two plants unless you have a big grow area.

*The *Past Medicinal Uses* information in this book is NOT intended to replace the advice of your licensed medical practitioner nor is it intended to diagnose, treat, cure or prevent any disease. It is merely a listing of some historical uses for this particular strain. Contact your licensed medical professional for advice on using various strains for your ailments.

Chocolope

Lineage: O.G. Chocolate Thai x Cantaloupe Haze

Genetics: 100% (S)

THC Content: 15% — 20%

Awards: 2007 Cannabis Cup 2nd Place
2008 Cannabis Cup 3rd Place
2010 Sativa Cup 2nd Place

Past Medicinal Uses*:

Anorexia
Arthritis Pain
Cancer
Chronic Pain
Depression
Glaucoma
Headache
Migraines
Nausea

Growing: Difficult

Flowering: 63 - 70 Days

Description: Chocolope is a pure Sativa strain. This plant will grow very tall. We would suggest topping the plant three to four times during the vegetative stage. You can find out more about topping your plants by visiting:

http://professorgrow.com/2010/09/21/mmj-corner-topping-and-cropping-to-increase-your-yield/

Also after two-to-two-and-a-half weeks in vegetative state (or when the plant is about 18 - 24 inches tall), change the light cycle to the flowering setting. The plant will continue to grow at least 12 to 18 inches in the flowering stage. Chocolope is not recommended for a small growing area.

*The *Past Medicinal Uses* information in this book is NOT intended to replace the advice of your licensed medical practitioner nor is it intended to diagnose, treat, cure or prevent any disease. It is merely a listing of some historical uses for this particular strain. Contact your licensed medical professional for advice on using various strains for your ailments.

Critical Mass

Lineage: Afghani x Skunk #1

Genetics: 60% (I)   40% (S)

THC Content: 19% — 22%

Awards: 2009 Spannabis Indoor Hydro Cup 1st Place

Past Medicinal Uses*:

Anorexia
Anxiety
Arthritis
Cancer
Chemotherapy
Chronic Pain
Insomnia
Migraines
Muscle Spasms

Growing: Moderate - Difficult

Flowering: 45 - 55 Days

Description: Critical Mass is an Indica dominant strain. It tends to grow to a medium height at full maturity.

We would only suggest topping this plant two to three times due to its extremely heavy yields. You will want to support the plants branches during its flowering stage because of the branches breaking off due to the weight of the flowering buds. You can find out more about topping your plants by visiting:

http://professorgrow.com/2010/09/21/mmj-corner-topping-and-cropping-to-increase-your-yield/

This plant will need a more room to grow because of its heavy yields. If you have a small grow room, we suggest you only grow two plants, more if you have a big grow area.

*The *Past Medicinal Uses* information in this book is NOT intended to replace the advice of your licensed medical practitioner nor is it intended to diagnose, treat, cure or prevent any disease. It is merely a listing of some historical uses for this particular strain. Contact your licensed medical professional for advice on using various strains for your ailments.

Diablo OG

Lineage: Chemdawg x Lemon Thai x Hindu Skunk

Genetics: 90% (I)   10% (S)

THC Content: 18% — 20%

Past Medicinal Uses*:

Anti-Inflammation
Anxiety
Appetite Stimulation
Arthritis
Chemotherapy
Chronic Pain
Insomnia
MS
Muscle Tension

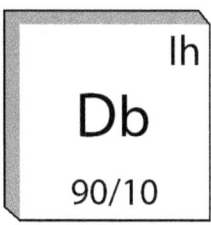

Growing: Moderate - Difficult

Flowering: 45 - 60 Days

Description: Diablo OG is an Indica dominant strain. This plant tends to grow to a small-medium height.

We suggest topping the plant one or two times during its vegetative stage to create more budding sites. You can find out more about topping your plants by visiting:

http://professorgrow.com/2010/09/21/mmj-corner-topping-and-cropping-to-increase-your-yield/

This strain tends to produce small to medium size buds as compared to other Indica strains. It also produces thinner branches which will need support during its flowering stage. This is a good plant to grow in small grow room. Three plants would do well in a smaller growing area.

*The *Past Medicinal Uses* information in this book is NOT intended to replace the advice of your licensed medical practitioner nor is it intended to diagnose, treat, cure or prevent any disease. It is merely a listing of some historical uses for this particular strain. Contact your licensed medical professional for advice on using various strains for your ailments.

Durban Poison

Lineage: South Africa Landrace Sativa

Genetics: 100% (S)

THC Content: 14% — 16%

Past Medicinal Uses*:

ADD
Anxiety
Appetite Stimulation
Cancer
Depression
Glaucoma
Inflammation
Joint Pain
Migraines
MS
Muscle Pain
Nausea
Neuropathic Pain
Phantom Limb Pain

Growing: Difficult

Flowering: 65 - 74 Days

Description: Durban Poison a pure Sativa strain. This plant will grow very tall. We would suggest topping the plant three to four times during the vegetative stage.

You can find out more about topping your plants by visiting: http://professorgrow.com/2010/09/21/mmj-corner-topping-and-cropping-to-increase-your-yield/

Also after two-or-two-and-a-half weeks in vegetative state (or when the plant is 18-to-24 inches tall), change the light cycle to the flowering setting. The plant will continue to grow at least 12 to 18 inches in the flowering stage, perhaps more.

Durban Poison is not recommended for a small growing area.

---

*The *Past Medicinal Uses* information in this book is NOT intended to replace the advice of your licensed medical practitioner nor is it intended to diagnose, treat, cure or prevent any disease. It is merely a listing of some historical uses for this particular strain. Contact your licensed medical professional for advice on using various strains for your ailments.

<u>Flo</u>

Lineage: Purple Thai x Afghan Indica

Genetics: 60% (S)   40% (I)

THC Content: 15% — 20%

Past Medicinal Uses*:

Anti-Anxiety
Anti-Fatigue
Appetite Stimulation
Energy
Mood Elevation
Muscle Tension
Pain Relief
Relaxation
Strong Ocular Attention

Growing: Moderate

Flowering: 50 - 55 Days

Description: Flo is a Sativa dominant strain, but the leaves and side-branching show off its Indica portion of the genetics. This plant tends to grow to a medium height. It also has good natural side branching.

This is a good plant for a small grow room. Two plants will do well. Just keep the side branching in mind when growing in a small grow room. It is a good plant to top two or three times to control the height. You can find out more about topping your plants by visiting:

http://professorgrow.com/2010/09/21/mmj-corner-topping-and-cropping-to-increase-your-yield/

*The *Past Medicinal Uses* information in this book is NOT intended to replace the advice of your licensed medical practitioner nor is it intended to diagnose, treat, cure or prevent any disease. It is merely a listing of some historical uses for this particular strain. Contact your licensed medical professional for advice on using various strains for your ailments.

G-13

Lineage: Afghani Descent

Genetics: 100% (I)

THC Content: 18% — 25%

Past Medicinal Uses*:

Arthritis
Cancer
Fibromyalgia
Glaucoma
HIV/AIDS
Inflammation
Joint Pain
Muscle Pain
Muscle Spasms
Nausea
Neuropathic Pain
Pain Killer
Skin Irritation

Growing: Easy

Flowering: 45 - 50 Days

Description: G-13 is a pure Indica strain. This plant tends to grow to a small-to-medium height. The G-13 plant doesn't produce any side branching. We would suggest topping this plant three to four times.

You can find out more about topping your plants by visiting:

http://professorgrow.com/2010/09/21/mmj-corner-topping-and-cropping-to-increase-your-yield/

This plant is good for a small grow room. Three plants could be grown well in a small grow area.

*The *Past Medicinal Uses* information in this book is NOT intended to replace the advice of your licensed medical practitioner nor is it intended to diagnose, treat, cure or prevent any disease. It is merely a listing of some historical uses for this particular strain. Contact your licensed medical professional for advice on using various strains for your ailments.

G-13 x Haze

Lineage: G-13 x Original Haze

Genetics: 80% (I)   20% (S)

THC Content: 18% — 20%

Awards: 2006 Cannabis Cup 2nd Place
2007 Cannabis Cup 1st Place

Past Medicinal Uses*:

ADD
ADHD
Anticonvulsant
Anxiety
Appetite Stimulation
Arthritis
Chronic Pain
Depression
Epilepsy
Gastrointestinal Disorder
Insomnia
Migraines

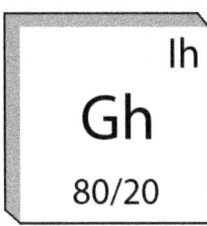

Growing: Moderate

Flowering: 55 - 70 Days

Description: G-13 crossed with Haze is an Indica dominant strain. It tends to grow to a medium height. It will have some side branching. We would still suggest to top the plant two to three times.

You can find out more about topping your plants by visiting:

http://professorgrow.com/2010/09/21/mmj-corner-topping-and-cropping-to-increase-your-yield/

This is a good plant for a small grow room. Two to three plants will do well. Keep in mind that it does have longer flowing time than most plants with this much Indica in them.

*The *Past Medicinal Uses* information in this book is NOT intended to replace the advice of your licensed medical practitioner nor is it intended to diagnose, treat, cure or prevent any disease. It is merely a listing of some historical uses for this particular strain. Contact your licensed medical professional for advice on using various strains for your ailments.

Girl Scout Cookies

Lineage: [OG Kush x Durban Poison] x [Grand Daddy Purple x OG Kush]

Genetics: 70% (S)   30% (I)

THC Content: 18% — 23%

Awards: 2013 First Place Hybrid LA Cannabis Cup

Past Medicinal Uses*:

AIDS
Anorexia
Anxiety
Arthritis
Depression
Epilepsy
Gastrointestinal Disorder
Glaucoma
Inflammation
Migraines
MS
Muscle Spasms
Nausea
Pain Relief
Parkinson's Disease
PTSD

Growing: Moderate - Difficult

Flowering: 65 - 75 Days

Description: Girl Scout Cookies is a Sativa dominant strain. This plant tends to grow to a medium-tall height.

We suggest topping the plant two to three times during the Vegetative stage to give it more side branching. You can find out more about topping your plants by visiting:

http://professorgrow.com/2010/09/21/mmj-corner-topping-and-cropping-to-increase-your-yield/

This plant can grow tall but not consistently. Keep this in mind if you have a small grow room. We suggest you only grow two or three plants for a small grow room. Use your best judgment for two or three plants.

*The *Past Medicinal Uses* information in this book is NOT intended to replace the advice of your licensed medical practitioner nor is it intended to diagnose, treat, cure or prevent any disease. It is merely a listing of some historical uses for this particular strain. Contact your licensed medical professional for advice on using various strains for your ailments.

God's Gift

Lineage: Grand Daddy Purple x OG Kush

Genetics: 90% (I)   10% (S)

THC Content: 18% - 19%

Past Medicinal Uses*:

Anorexia

Chemotherapy

Chronic Pain

Depression

Inflammation

Insomnia

MS

Nausea

Parkinson's Disease

Stress

Growing: Easy

Flowering: 55 - 65 Days

Description: God's Gift is an Indica dominant strain. This plant tends to grow to a small - medium height.

This is a great plant for new growers as it is very stress tolerant. We would suggest topping the plant two or three times during its vegetative stage to create more budding sites. You can find out more about topping your plants by visiting:

http://professorgrow.com/2010/09/21/mmj-corner-topping-and-cropping-to-increase-your-yield/

Three plants will do well in small grow room.

*The *Past Medicinal Uses* information in this book is NOT intended to replace the advice of your licensed medical practitioner nor is it intended to diagnose, treat, cure or prevent any disease. It is merely a listing of some historical uses for this particular strain. Contact your licensed medical professional for advice on using various strains for your ailments.

Grand Daddy Purple

Lineage: Purple Erkle x Big Bud

Genetics: 100% (I)

THC Content: 15% — 20%

Past Medicinal Uses*:

Appetite Stimulation
Insomnia
Joint Pain
Migraines
Nausea
Pain Relief
Relaxation

Growing: Easy

Flowering: 55 - 60 Days

Description: Grand Daddy Purple is a pure Indica strain. It tends to grow to a small-medium height. We would not suggest topping because of good natural side branching.

Because you don't have to spend time topping and otherwise controlling the height and shape of the plant, it is ideal for a beginning grower. The flowering time is shorter than many strains, as well.

This is a great plant for the small grow room. Three plants will do well in a small grow room.

*The *__Past Medicinal Uses__* information in this book is NOT intended to replace the advice of your licensed medical practitioner nor is it intended to diagnose, treat, cure or prevent any disease. It is merely a listing of some historical uses for this particular strain. Contact your licensed medical professional for advice on using various strains for your ailments.

Grapefruit

Lineage: Cinderella 99 (C99) x Fruity Sativa

Genetics: 80% (S)   20% (I)

THC Content: 15% — 20%

Past Medicinal Uses*:

Anti-Anxiety
Anti-Depression
Appetite Stimulation
Hypertension
Insomnia
Possible Ocular Relief
Relaxation

Growing: Easy-Moderate

Flowering: 50 - 55 Days

Description: Grapefruit is a Sativa dominant strain. It tends to grow to a small-medium height, shorter than most Sativa strains.

Like other Sativas, this is a fast growing plant. We would suggest topping the plant early on. Top it two to three times during the vegetative stage because of its fast growth. You can find out more about topping your plants by visiting:

http://professorgrow.com/2010/09/21/mmj-corner-topping-and-cropping-to-increase-your-yield/

This a good plant for a small grow room. Just keep an eye on it because of its quick growth, and fast flowering time.

*The *Past Medicinal Uses* information in this book is NOT intended to replace the advice of your licensed medical practitioner nor is it intended to diagnose, treat, cure or prevent any disease. It is merely a listing of some historical uses for this particular strain. Contact your licensed medical professional for advice on using various strains for your ailments.

Great White Shark

Lineage: Brazilian x South Indian x Super Skunk

Genetics: 50% (I)   50% (S)

THC Content: 14% — 20%

Awards: 1997 Cannabis Cup, 2nd Place
1997 Bio Cup, 2nd Place

Past Medicinal Uses*:

Anxiety
Appetite
Arthritis
Crohn's Disease
Chronic Pain
Cramps
Fibromyalgia
Gastrointestinal Disorder
Inflammation
Joint Pain
MS

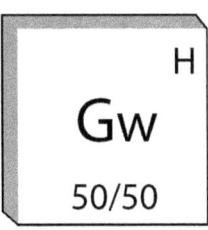

Growing: Easy

Flowering: 55 - 70 Days

Description: Great White Shark is a 50/50 mix of Indica and Sativa. This plant tends to grow to a medium-tall height and shows Indica dominant traits.

We would only suggest topping this plant two to three times due to its extremely heavy yields and good side branching. You can find out more about topping your plants by visiting:

http://professorgrow.com/2010/09/21/mmj-corner-topping-and-cropping-to-increase-your-yield/

You will want to support the plants branches during its flowering stage. This plant will need a more room to grow because of its heavy yields.

We suggest you only grow two plants unless you have a big grow area.

*The *Past Medicinal Uses* information in this book is NOT intended to replace the advice of your licensed medical practitioner nor is it intended to diagnose, treat, cure or prevent any disease. It is merely a listing of some historical uses for this particular strain. Contact your licensed medical professional for advice on using various strains for your ailments.

Hash Plant

Lineage: Lebanese x Thai x Northern Lights #1

Genetics: 100% (I)

THC Content: 15% — 20%

Past Medicinal Uses*:

Anxiety
Appetite Stimulation
Bipolar Disorder
Chronic Pain
Depression
Headache
Migraines
Nausea
Nervousness
Relaxation
Stress

Growing: Easy

Flowering: 40 - 45 Days

Description: Hash Plant is a pure Indica strain. It tends to grow small with not much side branching.

We would suggest topping three to four times to let the plant bush out a little more, because of its short stature. Bushing it out will produce more budding sites when it goes into flowering. You can find out more about topping your plants by visiting:

http://professorgrow.com/2010/09/21/mmj-corner-topping-and-cropping-to-increase-your-yield/

This plant is ideal for the small grow room. With heavy topping three plants will do well in a small grow room.

*The *Past Medicinal Uses* information in this book is NOT intended to replace the advice of your licensed medical practitioner nor is it intended to diagnose, treat, cure or prevent any disease. It is merely a listing of some historical uses for this particular strain. Contact your licensed medical professional for advice on using various strains for your ailments.

Hawaiian

Lineage: Hawaiian Landrace Sativa

Genetics: 100% (S)

THC Content: 15% — 20%

Past Medicinal Uses*:

Anxiety
Depression
Epilepsy
Glaucoma
Mood Elevation
Pain
Seizures

Growing: Difficult

Flowering: 86 - 100 Days

Description: Hawaiian a pure Sativa strain. This plant will grow very tall. It grows best outdoors where it has plenty of room.

We would suggest topping the plant three to four times during the vegetative stage. Find out more about topping your plants by visiting:

http://professorgrow.com/2010/09/21/mmj-corner-topping-and-cropping-to-increase-your-yield/

Because this plant grows so big, after two-to-two-and-a-half weeks in vegetative state (or when the plant reaches 18 to 24 inches tall), change the light cycle to the flowering setting. The plant will continue to grow at least 12 to 18 inches in the flowering stage. Hawaiian is not recommended for a small growing area.

*The *Past Medicinal Uses* information in this book is NOT intended to replace the advice of your licensed medical practitioner nor is it intended to diagnose, treat, cure or prevent any disease. It is merely a listing of some historical uses for this particular strain. Contact your licensed medical professional for advice on using various strains for your ailments.

Hawaiian Skunk

Lineage: Hawaiian x Skunk #1

Genetics: 60% (S)   40% (I)*

THC Content: 15% — 20%

Past Medicinal Uses*:

ADD
Anxiety
Appetite Stimulation
Depression
Insomnia
Migraines
Pain
Stress

Growing: Moderate

Flowering: 68 - 80 Days

Description: Hawaiian Skunk is a Sativa dominant strain. It tends to grow tall.

We would suggest topping the plant three to four times during the vegetative stage. You can find out more about topping your plants by visiting:

http://professorgrow.com/2010/09/21/mmj-corner-topping-and-cropping-to-increase-your-yield/

This plant will need a little more room to grow. Keep this in mind if you have a small grow room. We suggest you only grow two plants because of height, and long flowering time.

*The *Past Medicinal Uses* information in this book is NOT intended to replace the advice of your licensed medical practitioner nor is it intended to diagnose, treat, cure or prevent any disease. It is merely a listing of some historical uses for this particular strain. Contact your licensed medical professional for advice on using various strains for your ailments.

Hawaiian Snow

Lineage: Hawaiian Haze x Pure Haze x Neville's Haze

Genetics: 90% (S)   10% (I)

THC Content: 15% — 23%

Awards: 2003 Cannabis Cup 1st Place
2003 Sativa Cup 3rd Place

Past Medicinal Uses*:

Depression
Glaucoma
Inflammation
Joint Pain
Migraines
Nausea

Growing: Difficult

Flowering: 105 - 112 Days

Description: Hawaiian Snow is almost a pure Sativa. This plant will grow very tall. We would suggest topping the plant three to four times during the vegetative stage. You can find out more about topping your plants by visiting:

http://professorgrow.com/2010/09/21/mmj-corner-topping-and-cropping-to-increase-your-yield/

Also, after two-to-two-and-a-half weeks in vegetative state (or when the plant is 18-to-24 inches tall), change the light cycle to the flowering setting. The plant will continue to grow at least 12 to 18 inches in the flowering stage, perhaps more. Hawaiian Snow is not recommended for a small growing area.

*The *Past Medicinal Uses* information in this book is NOT intended to replace the advice of your licensed medical practitioner nor is it intended to diagnose, treat, cure or prevent any disease. It is merely a listing of some historical uses for this particular strain. Contact your licensed medical professional for advice on using various strains for your ailments.

Headband

Lineage: OG Kush x Master Kush x Sour Diesel]

Genetics: 85% (I)   15% (S)

THC Content: 15% — 20%

Awards: 2009 Cannabis Cup 3rd Place

Past Medicinal Uses*:

ADD
AIDS
Anorexia
Arthritis
Cancer
Depression
Epilepsy
Glaucoma
Migraines
Multiple Sclerosis
Muscle Spasms
Nausea
Pain
PTSD

Growing: Moderate

Flowering: 65 - 70 Days

Description: Headband is an Indica dominant strain. This plant tends to grow small-medium height. We would suggest topping two to three times during the vegetative stage.

You can find out more about topping your plants by visiting:

http://professorgrow.com/2010/09/21/mmj-corner-topping-and-cropping-to-increase-your-yield/

This is a good plant for a small grow room. Three plants would do well in a smaller growing area. Keep in mind that it does have a little longer flowering period than most Indicas.

*The *Past Medicinal Uses* information in this book is NOT intended to replace the advice of your licensed medical practitioner nor is it intended to diagnose, treat, cure or prevent any disease. It is merely a listing of some historical uses for this particular strain. Contact your licensed medical professional for advice on using various strains for your ailments.

Island Sweet Skunk

Lineage: BC Sweet Pink Grapefruit x Big Skunk #1

Genetics: 70% (S) 30% (I)

THC Content: 15% — 20%

Past Medicinal Uses*:

Appetite Stimulant
Chemotherapy
Epilepsy
Fatigue
Fibromyalgia
Glaucoma
Headaches
Mood Elevation
MS
Nausea
Nerve Pain
RSD (Reflex Sympathetic Dystrophy)

Growing: Moderate

Flowering: 50 - 55 Days

Description: Island Sweet Skunk is a Sativa dominant strain. This plant tends to grow tall. We would suggest topping three to four times during vegetative stage.

You can find out more about topping your plants by visiting:

http://professorgrow.com/2010/09/21/mmj-corner-topping-and-cropping-to-increase-your-yield/

This plant will need a little more room to grow. When putting in to your grow room, keep this in mind, especially if you have a small grow room. We suggest you only grow two plants because of height. The nice thing is that it has a shorter flowering time than most Sativas.

*The *Past Medicinal Uses* information in this book is NOT intended to replace the advice of your licensed medical practitioner nor is it intended to diagnose, treat, cure or prevent any disease. It is merely a listing of some historical uses for this particular strain. Contact your licensed medical professional for advice on using various strains for your ailments.

Jack Flash

Lineage: [Jack Herer x Super Skunk] x Haze

Genetics: 55% (I)   45% (S)

THC Content: 15% — 20%

Awards: 2004 Cannabis Cup 3rd Place

Past Medicinal Uses*:

Alzheimer's
Anxiety
Chronic Pain
Lou Gehrig's Disease
Muscle Spasms
Nausea
Tourette's

Growing: Moderate

Flowering: 60 - 70 Days

Description: Jack Flash is an Indica dominant strain. This plant tends to grow tall in height. This does not tend to produce many side branches. We would suggest topping the plant three to four times during the Vegetative stage. You can find out more about topping your plants by visiting:

http://professorgrow.com/2010/09/21/mmj-corner-topping-and-cropping-to-increase-your-yield/

After two-or-two-and-a-half weeks in Vegetative stage (or when the plant is 18-to-24 inches tall) , change the light cycle to the flowering setting.

The plant will continue to grow at least 12 to 18 inches in height and width during the flowering stage. This plant can be a little difficult because of its genetic crosses. Just be patient with the plant. We suggest you only grow two plants unless you have a big grow area.

*The *Past Medicinal Uses* information in this book is NOT intended to replace the advice of your licensed medical practitioner nor is it intended to diagnose, treat, cure or prevent any disease. It is merely a listing of some historical uses for this particular strain. Contact your licensed medical professional for advice on using various strains for your ailments.

Jack Herer

Lineage: Skunk #1 x NL #5 x Haze

Genetics: 80% (S)   20% (I)

THC Content: 15% — 20%

Awards: 1994 Cannabis Cup 1st Place
1999 Sativa Cup 1st Place

Past Medicinal Uses*:

ADD

Appetite

Energy

Fibromyalgia

Focus

Nervousness

Pain

Social Anxiety

Growing: Moderate-Difficult

Flowering: 60 - 70 Days

Description: Jack Herer is a Sativa dominant strain developed by the legendary Jack Herer. This plant tends to grow to a tall height. It has good natural side branching.

We would suggest topping the plant three to four times to control the height. You can find out more about topping your plants by visiting:

http://professorgrow.com/2010/09/21/mmj-corner-topping-and-cropping-to-increase-your-yield/

This plant can be a little difficult because of it's genetic crosses. Just be patient with the plant. We would not suggest this plant for a small grow room. It would do much better outdoors where it has plenty of room to grow.

*The *Past Medicinal Uses* information in this book is NOT intended to replace the advice of your licensed medical practitioner nor is it intended to diagnose, treat, cure or prevent any disease. It is merely a listing of some historical uses for this particular strain. Contact your licensed medical professional for advice on using various strains for your ailments.

Jilly Bean

Lineage: UKN Orange Skunk x Romulan x Cindy 99

Genetics: 70% (I)   30% (S)

THC Content: 12% — 18%

Past Medicinal Uses*:

Anxiety

Appetite Stimulation

Depression

Migraines

Nausea

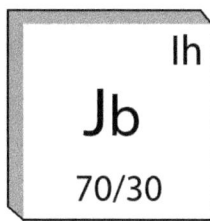

Growing: Easy-Moderate

Flowering: 56 - 63 Days

Description: Jilly Bean is an Indica dominant strain. (Note: We have seen it referenced as a Sativa dominant in some places, and we have seen Jilly Bean plants that looked more Sativa in their leaves and structure. Most sources referred to it as an Indica dominant.) This plant tends to grow small-medium height.

We would suggest topping three to four times during the vegetative stage, to let the plant bush out a little more and create more bud sites. You can find out more about topping your plants by visiting:

http://professorgrow.com/2010/09/21/mmj-corner-topping-and-cropping-to-increase-your-yield/

This plant is ideal for the small grow room. With heavy topping three plants will do well in a small grow room.

---

*The *Past Medicinal Uses* information in this book is NOT intended to replace the advice of your licensed medical practitioner nor is it intended to diagnose, treat, cure or prevent any disease. It is merely a listing of some historical uses for this particular strain. Contact your licensed medical professional for advice on using various strains for your ailments.

Juicy Fruit

Lineage: Afghani Landrace x Thai Landrace

Genetics: 75% (I)   25% (S)

THC Content: 15% — 20%

Past Medicinal Uses*:

Anti-Anxiety
Energy
Inspiration
Mild Body Pain Relief
Mood Elevation
Motivation
Ocular Attention

Growing: Easy-Moderate

Flowering: 56 - 63 Days

Description: Juicy Fruit is an Indica dominant strain. This plant tends to grow small-medium height.

We would suggest topping three to four times during the vegetative stage to let the plant bush out a little more and create more bud sites. You can find out more about topping your plants by visiting:

http://professorgrow.com/2010/09/21/mmj-corner-topping-and-cropping-to-increase-your-yield/

This plant is ideal for the small grow room. With heavy topping three plants will do well in a small grow room.

*The *Past Medicinal Uses* information in this book is NOT intended to replace the advice of your licensed medical practitioner nor is it intended to diagnose, treat, cure or prevent any disease. It is merely a listing of some historical uses for this particular strain. Contact your licensed medical professional for advice on using various strains for your ailments.

LA Confidential

Lineage: O.G. LA Affie x Afghani

Genetics: 100% (I)

THC Content: 15% — 20%

Awards: 2004 Indica Cup 3rd Place
2005 Indica Cup 2nd Place

Past Medicinal Uses*:

ADD
ADHD
Anxiety
Appetite Stimulation
Arthritis
Chronic Pain
Gastrointestinal Disorder
Insomnia
MS
Pain Relief

Growing: Easy

Flowering: 45 - 56 Days

Description: LA Confidential is a pure Indica strain. It tends to grow to a small to medium height. It also has nice side branching, creating multiple bud sites.

We would NOT suggest topping because LA Confidential already has good natural side branching.

This is a great plant for the small grow room. Three plants will do well in a small grow room.

This is a nice plant for a beginner to grow, as it doesn't get too big, doesn't need topping and has a fairly short flowering period.

*The *Past Medicinal Uses* information in this book is NOT intended to replace the advice of your licensed medical practitioner nor is it intended to diagnose, treat, cure or prevent any disease. It is merely a listing of some historical uses for this particular strain. Contact your licensed medical professional for advice on using various strains for your ailments.

Lambsbread

Lineage: Jamaican Landrace Sativa

Genetics: 100% (S)

THC Content: 15% — 18%

Past Medicinal Uses*:

ADD
ADHD
AIDS
Anorexia
Appetite
Arthritis
Cancer
Depression
Gastrointestinal Disorder
Glaucoma
Inflammation
Migraines
MS
Nausea
Pain (Day-Time)
PTSD

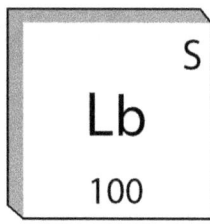

Growing: Easy - Moderate

Flowering: 60 - 70 Days

Description: Lambsbread is a pure Sativa strain. This plant will grow tall. We would suggest topping the plant three to four times during the Vegetative stage. You can find out more about topping your plants by visiting:

http://professorgrow.com/2010/09/21/mmj-corner-topping-and-cropping-to-increase-your-yield/

Also after two-or-two-and-a-half weeks in Vegetative stage (or when the plant is 18 to 24 inches tall), change the light cycle to the flowering setting. The plant will continue to grow at least 12 to 18 inches in height and width during the flowering stage, probably more.

Lambsbread in not recommended for a small growing area.

*The *Past Medicinal Uses* information in this book is NOT intended to replace the advice of your licensed medical practitioner nor is it intended to diagnose, treat, cure or prevent any disease. It is merely a listing of some historical uses for this particular strain. Contact your licensed medical professional for advice on using various strains for your ailments.

Lavender Kush

Lineage: Super Skunk x Big Skunk Korean x Afghani/Hawaiian

Genetics: 80% (I)   20% (S)

THC Content: 15% — 18%

Awards: 2005 Indica Cup 1st Place

Past Medicinal Uses*:

Anxiety
Appetite
Bipolar Disorder
Insomnia
Migraines
Nausea
Pain Relief
PTSD

The 65 Most Common Strains

Growing: Easy

Flowering: 53 - 63 Days

Description: Lavender Kush is an Indica dominant strain. This plant tends to grow to a medium height.

We would suggest topping two to three times during the vegetative stage. This will help keep the plant at a good, manageable height and increase the number of bud sites. You can find out more about topping your plants by visiting:

http://professorgrow.com/2010/09/21/mmj-corner-topping-and-cropping-to-increase-your-yield/

Three plants will do well in a small grow room.

*The *Past Medicinal Uses* information in this book is NOT intended to replace the advice of your licensed medical practitioner nor is it intended to diagnose, treat, cure or prevent any disease. It is merely a listing of some historical uses for this particular strain. Contact your licensed medical professional for advice on using various strains for your ailments.

Lemon Skunk

Lineage: Skunk #1 x Citral

Genetics: 60% (S)   40% (I)

THC Content: 18% — 22%

Past Medicinal Uses*:

ADHD
Appetite
Autism
Cancer
Epilepsy
Insomnia
Muscle Pain
Seizures

Growing: Moderate

Flowering: 50 - 56 Days

Description: Lemon Skunk is a Sativa dominant strain. This plant tends to grow to a medium-tall height.

We would suggest topping three to four times during vegetative stage. This will help keep the plant from getting too tall. You can find out more about topping your plants by visiting:

http://professorgrow.com/2010/09/21/mmj-corner-topping-and-cropping-to-increase-your-yield/

This plant needs a little more room to grow. When putting in to your grow room keep this in mind, especially if you have a small grow room. We suggest you only grow two plants unless you have a big grow area.

*The *Past Medicinal Uses* information in this book is NOT intended to replace the advice of your licensed medical practitioner nor is it intended to diagnose, treat, cure or prevent any disease. It is merely a listing of some historical uses for this particular strain. Contact your licensed medical professional for advice on using various strains for your ailments.

Mango Haze

Lineage: [NL #5 x Skunk] x Haze

Genetics: 50% (S)   50% (I)

THC Content: 15% — 20%

Past Medicinal Uses*:

Anxiety
Depression
Eating Disorders
Inflammation
Migraines
Nausea
Pain Relief

Growing: Moderate-Difficult

Flowering: 50 - 60 Days

Description: Mango Haze strain is a 50/50 mix of Indica and Sativa. This plant tends to grow to a medium-tall height.

We would suggest topping three to four times during vegetative stage. This will help keep the plant at a good height. You can find out more about topping your plants by visiting:

http://professorgrow.com/2010/09/21/mmj-corner-topping-and-cropping-to-increase-your-yield/

This plant will need a little more room to grow. Keep this in mind if you have a small grow room. We suggest you only grow two plants unless you have a big grow area.

---

*The *__Past Medicinal Uses__* information in this book is NOT intended to replace the advice of your licensed medical practitioner nor is it intended to diagnose, treat, cure or prevent any disease. It is merely a listing of some historical uses for this particular strain. Contact your licensed medical professional for advice on using various strains for your ailments.

Master Kush

Lineage: Hindu Kush x Skunk

Genetics: 80% (I)   20% (S)

THC Content: 15% — 18%

Past Medicinal Uses*:

Anxiety
Asthma
Creative Thoughts
Energy
Increased Appetite
Nausea
Relaxation

Growing: Easy

Flowering: 63 - 70 Days

Description: Master Kush is an Indica dominant strain. This plant tends to grow small.

We would suggest topping two to three times during the vegetative stage to increase the number of bud sites. You can find out more about topping your plants by visiting:

http://professorgrow.com/2010/09/21/mmj-corner-topping-and-cropping-to-increase-your-yield/

This is a good plant for a small grow room. Three plants would do well in a smaller growing area. Just keep in mind that it does have a little longer flowering period than most Indicas.

*The *Past Medicinal Uses* information in this book is NOT intended to replace the advice of your licensed medical practitioner nor is it intended to diagnose, treat, cure or prevent any disease. It is merely a listing of some historical uses for this particular strain. Contact your licensed medical professional for advice on using various strains for your ailments.

Maui Wowie

Lineage: Hawaiian Hybrid

Genetics: 70% (S)   30% (I)

THC Content: 8% — 15%

Past Medicinal Uses*:

Arthritis
Asthma
Bipolar Disorder
Depression
Glaucoma
Migraines
PTSD
Tourette's Syndrome

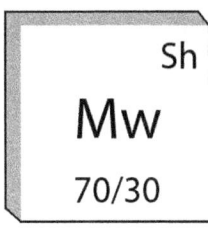

Growing: Moderate-Difficult

Flowering: 60 - 75 Days

Description: Maui Wowie is a Sativa dominant strain. This plant tends to grow to a medium-tall height.

We would suggest topping three to four times during vegetative stage. This will help keep the plant at a good height. You can find out more about topping your plants by visiting:

http://professorgrow.com/2010/09/21/mmj-corner-topping-and-cropping-to-increase-your-yield/

This plant will need a little more room to grow. When putting in to your grow room keep this in mind. If you have a small grow room. We suggest you only grow two plants unless you have a big grow area.

*The _Past Medicinal Uses_ information in this book is NOT intended to replace the advice of your licensed medical practitioner nor is it intended to diagnose, treat, cure or prevent any disease. It is merely a listing of some historical uses for this particular strain. Contact your licensed medical professional for advice on using various strains for your ailments.

Northern Lights

Lineage: Afghan x Skunk #1 x Haze

Genetics: 95% (I) 5% (S)

THC Content: 15% — 20%

Awards: 1993 Cannabis Cup 2nd Place

Past Medicinal Uses*:

Anxiety

Depression

Hypertension

Insomnia

Lower Back Pain

Nausea

PTSD

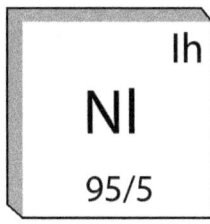

Growing: Easy

Flowering: 45 - 50 Days

Description: Northern Lights is almost a pure Indica strain. This plant tends to grow small- medium height.

We would suggest topping two to three times during the vegetative stage to help produce more budding sites.

You can find out more about topping your plants by visiting:

http://professorgrow.com/2010/09/21/mmj-corner-topping-and-cropping-to-increase-your-yield/

This is a good plant for a small grow room. Three plants would do well in a smaller growing area.

*The *Past Medicinal Uses* information in this book is NOT intended to replace the advice of your licensed medical practitioner nor is it intended to diagnose, treat, cure or prevent any disease. It is merely a listing of some historical uses for this particular strain. Contact your licensed medical professional for advice on using various strains for your ailments.

NYC Diesel

Lineage: Sour Diesel x Afghani x Hawaiian

Genetics: 60% (S)   40% (I)

THC Content: 15% — 20%

Awards: 2001 Sativa Cup 3rd Place
2002 Cannabis Cup 2nd Place
2002 Sativa Cup 2nd Place
2003 Cannabis Cup 2nd Place

Past Medicinal Uses*:

Appetite
Depression
Migraines
MS
Stomach Ailments

Growing: Moderate

Flowering: 55 - 75 Days

Description: New York City Diesel is a Sativa dominant strain. It tends to grow to a medium height.

It will have some side branching. We would still suggest you top the plant two to three times during the vegetative stage to increase side branching and budding sites. You can find out more about topping your plants by visiting:

http://professorgrow.com/2010/09/21/mmj-corner-topping-and-cropping-to-increase-your-yield/

This is a good plant for a small grow room. Two to three plants will do well in a small space. Keep in mind that it has a longer flowering time.

---

*The *Past Medicinal Uses* information in this book is NOT intended to replace the advice of your licensed medical practitioner nor is it intended to diagnose, treat, cure or prevent any disease. It is merely a listing of some historical uses for this particular strain. Contact your licensed medical professional for advice on using various strains for your ailments.

OG Kush

Lineage: Sour Diesel x Chem Dawg

Genetics: 85% (S)   15% (I)

THC Content: 15% — 20%

Past Medicinal Uses*:

Anti-Anxiety
Anti-Depression
Appetite Stimulation
Headaches
Minor Stomach Discomfort
Mood Elevation
Strong Ocular Attention

The 65 Most Common Strains

Growing: Moderate- Difficult

Flowering: 55 - 65 Days

Description: OG Kush is a Sativa dominant strain. It tends to grow medium-tall.

We would suggest topping the plant three to four times during the vegetative stage to give it more side branching. You can find out more about topping your plants by visiting:

http://professorgrow.com/2010/09/21/mmj-corner-topping-and-cropping-to-increase-your-yield/

This plant will need a little more room to grow. When putting in to your grow room keep this in mind. If you have a small grow room. We suggest you only grow two plants because of height, and its potential to take up space.

---

*The *Past Medicinal Uses* information in this book is NOT intended to replace the advice of your licensed medical practitioner nor is it intended to diagnose, treat, cure or prevent any disease. It is merely a listing of some historical uses for this particular strain. Contact your licensed medical professional for advice on using various strains for your ailments.

Purple Haze

Lineage: [Columbian Haze x Meao Thai] x Dutch Skunk

Genetics: 100% (S)†

THC Content: 18% — 20%

Past Medicinal Uses*:

ADD
ADHD
Anxiety
Appetite Stimulation
Bipolar Disorder
Cancer
Cramps
Depression
Glaucoma
HIV/AIDS
Insomnia
Joint Pain
Muscle Pain
Muscle Spasms
PMDD
PMS
PTSD

Growing: Difficult

Flowering: 50 - 65 Days

Description: Purple Haze is a pure Sativa strain. This plant will grow very tall. We would suggest topping the plant three to four times during the vegetative stage to keep the height under control. You can find out more about topping your plants by visiting:

http://professorgrow.com/2010/09/21/mmj-corner-topping-and-cropping-to-increase-your-yield/

Also after two-to-two-and-a-half weeks in vegetative state (or when the plant is 18-to-24 inches tall), change the light cycle to the flowering setting. The plant will continue to grow at least 12 to 18 inches in the flowering stage. Purple Haze is not recommended for a small growing area.

†We have seen Purple Kush listed as 100% Sativa and as 85% Sativa, depending on the breeder.

*The *Past Medicinal Uses* information in this book is NOT intended to replace the advice of your licensed medical practitioner nor is it intended to diagnose, treat, cure or prevent any disease. It is merely a listing of some historical uses for this particular strain. Contact your licensed medical professional for advice on using various strains for your ailments.

Shiva Shanti

Lineage: Northern Lights #5 x Skunk #1

Genetics: 85% (I)   15% (S)

THC Content: 15% — 20%

Awards: 2003 Outdoor Cup 1st Place

Past Medicinal Uses*:

ADHD
Anxiety
Appetite Stimulation
Arthritis
Chronic Pain
Epilepsy
Glaucoma
Insomnia
Migraines
Muscle Spasms
Nausea
Pain Relief
Panic Attacks
PMS
Relaxation

Growing: Easy

Flowering: 44 - 60 Days

Description: Shiva Shanti is an Indica dominant strain. This plant tends to grow small to medium in height.

We would suggest topping two to three times during the vegetative stage to increase side branching and budding sites. You can find out more about topping your plants by visiting:

http://professorgrow.com/2010/09/21/mmj-corner-topping-and-cropping-to-increase-your-yield/

This is a good plant for a small grow room. Three plants would do well in a smaller growing area.

*The *Past Medicinal Uses* information in this book is NOT intended to replace the advice of your licensed medical practitioner nor is it intended to diagnose, treat, cure or prevent any disease. It is merely a listing of some historical uses for this particular strain. Contact your licensed medical professional for advice on using various strains for your ailments.

Silver Haze

Lineage: Skunk x [NL x Haze]

Genetics: 50% (S)   50% (I)

THC Content: 15% — 18%

Awards: 2005 Cannabis Cup 3rd Place

Past Medicinal Uses*:

Anxiety
Appetite
Epilepsy
Multiple Sclerosis

Growing: Easy

Flowering: 56 - 70 Days

Description: Silver Haze is a 50/50 hybrid. This plant tends to grow to a medium-tall height.

We would suggest topping three to four times during vegetative stage. This will help keep the plant at a good height. You can find out more about topping your plants by visiting:

http://professorgrow.com/2010/09/21/mmj-corner-topping-and-cropping-to-increase-your-yield/

This plant needs a little more room to grow. Keep this in mind if you have a small grow room. We suggest you only grow two plants unless you have a big grow area.

*The *Past Medicinal Uses* information in this book is NOT intended to replace the advice of your licensed medical practitioner nor is it intended to diagnose, treat, cure or prevent any disease. It is merely a listing of some historical uses for this particular strain. Contact your licensed medical professional for advice on using various strains for your ailments.

Skunk

Lineage: [Afghani x Acapulco Gold] x Columbian Gold

Genetics: 60% (I) 40% (S)

THC Content: 10% — 15%

Awards: 1988 Cannabis Cup 1st Place
1989 Indica Cup 2nd Place

Past Medicinal Uses*:

ADD
Alzheimer's Disease
Anxiety
Appetite
Depression
Epilepsy
Insomnia
Nausea
Pain
Tension

Growing: Easy-Moderate

Flowering: 60 - 74 Days

Description: Skunk is an Indica dominant strain. This plant tends to grow to a medium height.

We would suggest topping two to three times. This will help keep the plant at a good height. You can find out more about topping your plants by visiting:

http://professorgrow.com/2010/09/21/mmj-corner-topping-and-cropping-to-increase-your-yield/

Three plants will do well in a small grow room. Just keep in mind that it does have a little longer flowering period than most Indicas.

---

*The *Past Medicinal Uses* information in this book is NOT intended to replace the advice of your licensed medical practitioner nor is it intended to diagnose, treat, cure or prevent any disease. It is merely a listing of some historical uses for this particular strain. Contact your licensed medical professional for advice on using various strains for your ailments.

Skywalker

Lineage: Mazar x Blueberry

Genetics: 65% (I)   35% (S)

THC Content: 10% — 18%

Past Medicinal Uses*:

Anxiety

Chronic Pain

Insomnia

Muscle Pain

Nausea

Pain

Growing: Easy

Flowering: 60 - 74 Days

Description: Skywalker is an Indica dominant strain. This plant tends to grow to a medium height.

We would suggest topping two to three times during the vegetative stage. This will help keep the plant at a good height, encourage side branching and increase budding sites. You can find out more about topping your plants by visiting:

http://professorgrow.com/2010/09/21/mmj-corner-topping-and-cropping-to-increase-your-yield/

Three plants should do well in a small grow room if you do your topping two or three times.

*The *Past Medicinal Uses* information in this book is NOT intended to replace the advice of your licensed medical practitioner nor is it intended to diagnose, treat, cure or prevent any disease. It is merely a listing of some historical uses for this particular strain. Contact your licensed medical professional for advice on using various strains for your ailments.

Sour Diesel

Lineage: Mexican Sativa x Chemo

Genetics: 90% (S)   10% (I)

THC Content: 12% — 15%

Past Medicinal Uses*:

ADD
ADHD
Anxiety
Depression
Edema
Epilepsy
Fibromyalgia
Migraines
Nausea
Radiculopathy

Growing: Moderate

Flowering: 75 - 80 Days

Description: Sour Diesel is almost a pure Sativa strain. This plant will grow very tall. We would suggest topping the plant three to four times during the vegetative stage to control the height. You can find out more about topping your plants by visiting:

http://professorgrow.com/2010/09/21/mmj-corner-topping-and-cropping-to-increase-your-yield/

After two-to-two-and-a-half weeks in vegetative state (or when the plant is 18-to-24 inches tall), change the light cycle to the flowering setting. The plant will continue to grow at least 12 to 18 inches in the flowering stage.

Sour Diesel is not recommended for a small growing area.

*The *Past Medicinal Uses* information in this book is NOT intended to replace the advice of your licensed medical practitioner nor is it intended to diagnose, treat, cure or prevent any disease. It is merely a listing of some historical uses for this particular strain. Contact your licensed medical professional for advice on using various strains for your ailments.

Space Queen (AKA Space Jill)

Lineage: Romulan X Cinderella 99

Genetics: 70% (S)   30% (I)

THC Content: 14% — 20%

Past Medicinal Uses*:

ADD

Alzheimer's

Anti-Convulsant

Arthritis

Chronic Pain

Depression

Glaucoma

Migraines

Muscle Spasms

Muscle Tension

Nausea

Pain Relief

The 65 Most Common Strains

Growing: Moderate

Flowering: 50 - 60 Days

Description: Space Queen (AKA Space Jill) is a Sativa dominant strain. This plant tends to grow to a good medium height with lots of natural side branching.

We would suggest topping the plant two or three times during the Vegetative stage to control height and yield. You can find out more about topping your plants by visiting:

http://professorgrow.com/2010/09/21/mmj-corner-topping-and-cropping-to-increase-your-yield/

Because of such good side branching we suggest to stake or tie up branches for support during the flowering stage. This plant will need a more room to grow. Keep this in mind if you have a small grow room. We suggest you only grow two plants unless you have a big grow area.

*The *Past Medicinal Uses* information in this book is NOT intended to replace the advice of your licensed medical practitioner nor is it intended to diagnose, treat, cure or prevent any disease. It is merely a listing of some historical uses for this particular strain. Contact your licensed medical professional for advice on using various strains for your ailments.

Strawberry Cough

Lineage: Strawberry Fields x Haze

Genetics: 75% (S)   25% (I)

THC Content: 15% — 20%

Past Medicinal Uses*:

ADD
Back Pain
Depression
Gastroenteritis
Joint Pain
Migraines
PTSD

Growing: Easy-Moderate

Flowering: 56 - 63 Days

Description: Strawberry Cough is a Sativa dominant strain. It tends to grow to a medium- tall height.

The good thing about this strain is the shorter flowering time for a Sativa dominant strain. And, unlike most Sativas, this strain can be grown in a small grow room.

We suggest two plants. This is because of topping the plants during the vegetative cycle. Topping the plant three to four time will do fine. You can find out more about topping your plants by visiting:

http://professorgrow.com/2010/09/21/mmj-corner-topping-and-cropping-to-increase-your-yield/

*The *Past Medicinal Uses* information in this book is NOT intended to replace the advice of your licensed medical practitioner nor is it intended to diagnose, treat, cure or prevent any disease. It is merely a listing of some historical uses for this particular strain. Contact your licensed medical professional for advice on using various strains for your ailments.

Strawberry Diesel

Lineage: NYC Diesel x Strawberry Cough

Genetics: 50% (S)   50% (I)

THC Content: 15% — 20%

Past Medicinal Uses*:

ADHD
Anti-Convulsants
Anti-Inflammatory
Anxiety
Appetite Stimulant
Arthritis
Chronic Pain
Glaucoma
Nausea
Stress and Tension

The 65 Most Common Strains

Growing: Moderate

Flowering: 55 - 60 Days

Description: Strawberry Diesel is a 50/50 mix of Indica and Sativa. This plant tends to grow to a medium-tall height.

We suggest topping the plant two or three times during its vegetative stage to create more budding sites and to control height. You can find out more about topping your plants by visiting:

http://professorgrow.com/2010/09/21/mmj-corner-topping-and-cropping-to-increase-your-yield/

This plant can grow tall but not consistently. Keep this in mind if you have a small grow room. We suggest you only grow two - three plants for a small grow room. Use your best judgment for whether to grow two or three plants.

---

*The *Past Medicinal Uses* information in this book is NOT intended to replace the advice of your licensed medical practitioner nor is it intended to diagnose, treat, cure or prevent any disease. It is merely a listing of some historical uses for this particular strain. Contact your licensed medical professional for advice on using various strains for your ailments.

Super Lemon Haze

Lineage: Skunk x NL x Haze

Genetics: 75% (S)   25% (I)

THC Content: 18% — 22%

Awards: 2009 Cannabis Cup 1st Place
2009 Sativa Cup 2nd Place
2010 Cannabis Cup 2nd Place

Past Medicinal Uses*:

ADD

ADHD

Appetite Stimulation

Attentive Mind set

Focus

Possible Ocular Relief

Growing: Moderate

Flowering: 67 - 85 Days

Description: Super Lemon Haze is a Sativa dominant strain. It tends to grow to a medium-tall height.

We would suggest topping three to four times during vegetative stage. This will help keep the plant at a good height. You can find out more about topping your plants by visiting:

http://professorgrow.com/2010/09/21/mmj-corner-topping-and-cropping-to-increase-your-yield/

This plant will need a little more room to grow. When putting Super Lemon Haze into your grow room keep this in mind, especially if you have a small grow room. We suggest you only grow two plants unless you have a big grow area.

*The *Past Medicinal Uses* information in this book is NOT intended to replace the advice of your licensed medical practitioner nor is it intended to diagnose, treat, cure or prevent any disease. It is merely a listing of some historical uses for this particular strain. Contact your licensed medical professional for advice on using various strains for your ailments.

Super Silver Haze

Lineage: Skunk x NL x Haze

Genetics: 75% (S)   25% (I)

THC Content: 15% — 20%

Awards: 1998 Cannabis Cup 1st Place
1999 Cannabis Cup 1st Place
2007 Cannabis Cup 3rd Place
2008 Cannabis Cup 1st Place

Past Medicinal Uses*:

ADD
ADHD
Anxiety
Appetite
Back Pain
Cramps
Depression
Glaucoma
Joint Pain
Migraines
Muscle Pain
PMDD
PMS

The 65 Most Common Strains

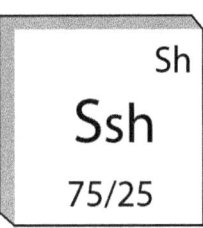

Growing: Moderate

Flowering: 56 - 70 Days

Description: Super Silver Haze is a Sativa dominant strain. Super Silver Haze tends to grow to a medium-tall height.

We would suggest topping three to four times during vegetative stage. This will help keep the plant at a good height and increase side branching. You can find out more about topping your plants by visiting:

http://professorgrow.com/2010/09/21/mmj-corner-topping-and-cropping-to-increase-your-yield/

This plant needs a little more room to grow. When putting into your grow room keep this in mind, if you have a small grow room. We suggest you only grow two plants unless you have a big growing area.

*The *Past Medicinal Uses* information in this book is NOT intended to replace the advice of your licensed medical practitioner nor is it intended to diagnose, treat, cure or prevent any disease. It is merely a listing of some historical uses for this particular strain. Contact your licensed medical professional for advice on using various strains for your ailments.

Sweet Tooth

Lineage: [Pink Grapefruit X Blueberry] X Grapefruit

Genetics: 90% (I)   10% (S)

THC Content: 15% — 20%

Awards: 2001 Cannabis Cup, 1st Place

Past Medicinal Uses*:

Anti-Convulsant
Anxiety
Appetite
Depression
Insomnia
Pain Relief
PTSD
Tension

Growing: Easy

Flowering: 40 - 50 Days

Description: Sweet Tooth is an Indica dominant strain. This plant tends to grow to a small-medium height.

We would suggest topping the plant two to three times during the Vegetative stage to give it more side branching. You can find out more about topping your plants by visiting:

http://professorgrow.com/2010/09/21/mmj-corner-topping-and-cropping-to-increase-your-yield/

This is a good plant to grow in small grow room. Three plants would do well in a smaller growing area.

*The *Past Medicinal Uses* information in this book is NOT intended to replace the advice of your licensed medical practitioner nor is it intended to diagnose, treat, cure or prevent any disease. It is merely a listing of some historical uses for this particular strain. Contact your licensed medical professional for advice on using various strains for your ailments.

Train Wreck

Lineage: Afghani x Thai x [Mexican x Columbian]

Genetics: 90% (S)   10% (I)

THC Content: 15% — 20%

Past Medicinal Uses*:

Anxiety
Appetite
Arthritis
Diabetic Neuropathy
Insomnia

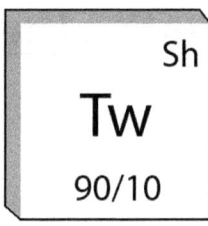

Growing: Moderate- Difficult

Flowering: 60 - 70 Days

Description: Train Wreck strain is almost a pure Sativa strain. It tends to grow to a medium-tall height.

We would suggest topping three to four times during vegetative stage. This will help keep the plant at a good height and increase side branching. You can find out more about topping your plants by visiting:

http://professorgrow.com/2010/09/21/mmj-corner-topping-and-cropping-to-increase-your-yield/

This plant will need a little more room to grow. When putting in to your grow room keep this in mind, if you have a small grow room. We suggest you only grow two plants unless you have a big grow area.

*The *Past Medicinal Uses* information in this book is NOT intended to replace the advice of your licensed medical practitioner nor is it intended to diagnose, treat, cure or prevent any disease. It is merely a listing of some historical uses for this particular strain. Contact your licensed medical professional for advice on using various strains for your ailments.

Vanilla Kush

Lineage: Kashmir x Afghan Kush

Genetics: 85% (I)   15% (S)

THC Content: 20% — 22%

Awards: 2009 Cannabis Cup 2nd Place

Past Medicinal Uses*:

Anxiety
Appetite
Back Pain
Cramps
Depression
Joint Pain
Migraines
Muscle Pain
PMDD
PMS

The 65 Most Common Strains

Growing: Easy

Flowering: 60 - 65 Days

Description: Vanilla Kush is an Indica dominant strain. This plant tends to grow small, only about 1-1/2 feet to 2 feet tall.

We would suggest you don't top the plant because this strain has very good natural side branching. Topping this strain will only decrease your yield. The yields can be excellent for such a short plant.

This is a good plant to grow in small grow room. Three plants would do well in a smaller growing area, as long as you can make room for the side branching.

*The *Past Medicinal Uses* information in this book is NOT intended to replace the advice of your licensed medical practitioner nor is it intended to diagnose, treat, cure or prevent any disease. It is merely a listing of some historical uses for this particular strain. Contact your licensed medical professional for advice on using various strains for your ailments.

White Rhino

Lineage: Afghan x Brazilian x South Indian

Genetics: 60 (I)   40% (S)

THC Content: 15% — 20%

Past Medicinal Uses*:

Anti-Inflammatory
Anxiety
Appetite
Chronic Pain
Insomnia
Migraines
MS
Muscle Tension
Nausea
Stomach Pain
Stress

Growing: Easy

Flowering: 60 - 74 Days

Description: White Rhino is an Indica dominant strain. This plant tends to grow small-medium height.

We would suggest topping two to three times during the vegetative stage to increase budding sites. You can find out more about topping your plants by visiting:

http://professorgrow.com/2010/09/21/mmj-corner-topping-and-cropping-to-increase-your-yield/

This is a good plant for a small grow room. Three plants would do well in a smaller growing area. Just keep in mind that White Rhino does have a little longer flowering period than most Indicas.

*The *Past Medicinal Uses* information in this book is NOT intended to replace the advice of your licensed medical practitioner nor is it intended to diagnose, treat, cure or prevent any disease. It is merely a listing of some historical uses for this particular strain. Contact your licensed medical professional for advice on using various strains for your ailments.

White Widow

Lineage: Brazilian x South Indian

Genetics: 60% (S) 40% (I)

THC Content: 15% — 20%

Awards: 1995 Cannabis Cup 1st Place

Past Medicinal Uses*:

Anxiety
Appetite Stimulation
Cachexia
Cancer
Depression
Epilepsy
Fibromyalgia
Glaucoma
Hepatitis C
HIV/AIDS
PTSD
Seizures

Growing: Easy

Flowering: 60 - 74 Days

Description: White Widow is a Sativa dominant strain. This plant tends to grow small-medium height.

We would suggest topping two to three times during the vegetative stage to increase budding sites. You can find out more about topping your plants by visiting:

http://professorgrow.com/2010/09/21/mmj-corner-topping-and-cropping-to-increase-your-yield/

This is a good plant for a small grow room. Three plants would do well in a smaller growing area.

---

*The *Past Medicinal Uses* information in this book is NOT intended to replace the advice of your licensed medical practitioner nor is it intended to diagnose, treat, cure or prevent any disease. It is merely a listing of some historical uses for this particular strain. Contact your licensed medical professional for advice on using various strains for your ailments.

Marijuana Laws

In the previous edition, we listed the medical marijuana laws for each state that allowed it. Because of the very dynamic nature of laws related to marijuana, the information was quickly out-of-date.

Currently, 19 states plus Washington, DC have passed Medical Marijuana Laws. States differ widely in their rules. Some states allow home growing, while others do not. Some states allow dispensaries, others license growers and some are setting up state owned stores to sell marijuana to patients. Laws are constantly changing as states update their rules and more pass new laws.

Colorado and Washington have approved recreational marijuana and several others are considering it. Some have decriminalized possession.

Instead of listing info that will soon be out of date, we want to refer you to places where you can easily get the latest info on various states. The following organizations are keeping track of the law changes:

NORML
http://norml.org

Marijuana Policy Project
http://www.mpp.org

Americans for Safe Access
http://safeaccessnow.org

Toke of The Town
http://www.tokeofthetown.com

Ailment Index

This Index lists Ailments (alphabetically) along with the cannabis strains that have been used in the past by patients seeking relief for those Ailments.

The information in this index is **NOT** intended to replace the advice of your licensed medical practitioner nor is it intended to diagnose, treat, cure or prevent any disease. It is merely a listing of some historical uses for this particular strain. Contact your licensed medical professional for advice on using various strains of cannabis for your ailments.

As a reference, you can look up an ailment in this index and see what strains have historically been used for that ailment. The Strains are all listed in the Table of Contents (alphabetically).

The authors wish to remind the readers that the use of cannabis for medical purposes is only approved in a limited number of states. We advise you to check your state's laws and regulations and comply with them if you are seeking to be a medical marijuana patient. Use of cannabis for any purpose, including medicinal, remains unlawful under Federal laws.

Ailments and Strains

As a reference, you can look up an ailment in this index and see what strains have historically been used for that ailment. The authors make no recommendations regarding the appropriateness of using any cannabis strain for any purpose.

Aches
AK-47
Blackberry Kush
Bubba Kush
Buddha's Sister
Chocolope
Hash Plant
Island Sweet Skunk
OG Kush

ADD
Bubble Berry
Bubble Gum
Buddha's Sister
Champagne
Durban Poison
G-13 x Haze
Hawaiian Skunk
Headband
Jack Herer
LA Confidential
Lambsbread
Purple Haze
Skunk

ADD (cont.)
Sour Diesel
Space Queen
Strawberry Cough
Super Lemon Haze
Super Silver Haze

ADHD
Buddha's Sister
Champagne
G-13 x Haze
LA Confidential
Lambsbread
Lemon Skunk
Purple Haze
Shiva Shanti
Sour Diesel
Strawberry Diesel
Super Lemon Haze
Super Silver Haze

AIDS
Afghan Goo (aka Afghooey)
Blueberry

AIDS (cont.)
Blue Dream
Champagne
G-13
Girl Scout Cookies
Headband
Lambsbread
Purple Haze
White Widow

Alcoholism
Afghan Goo (aka Afghooey)

ALS
Buddha's Sister
Jack Flash

Alzheimer's Disease
AK-47
Bubble Gum
Jack Flash
Skunk
Space Queen

Anorexia
Afghan Goo (aka Afghooey)
Blueberry
Blue Dream
Blue Moonshine
Chocolope
Girl Scout Cookies

Anorexia (cont.)
God's Gift
Headband
Lambsbread

Anti-Anxiety
Afghan Goo (aka Afghooey)
Big Bud
Blackberry Kush
Blue Moonshine
Blueberry
Bruce Banner
Bubba Kush
Bubble Gum
Champagne
Chem Dog (aka Chem Dawg)
Diablo OG
Durban Poison
Flo
G-13 x Haze
Girl Scout Cookies
Grapefruit
Great White Shark
Hash Plant
Hawaiian
Hawaiian Skunk
Jack Flash
Jack Herer
Jilly Bean
Juicy Fruit
LA Confidential

Ailments & Strains

Anti-Anxiety (cont.)

    Lavender Kush
    Mango Haze
    Master Kush
    Northern Lights
    OG Kush
    Purple Haze
    Shiva Shanti
    Silver Haze
    Skunk
    Skywalker
    Sour Diesel
    Strawberry Diesel
    Super Silver Haze
    Sweet Tooth
    Train Wreck
    Vanilla Kush
    White Rhino
    White Widow

Anti-Convulsant

    Bubble Gum
    Blue Moonshine
    G-13 x Haze
    Space Queen
    Strawberry Diesel
    Sweet Tooth

Anti-Depression

    Afghan Goo (aka Afghooey)
    Afghooey Train Wreck

Anti-Depression (cont.)

    AK-47
    Big Bud
    Blue Dream
    Bubble Berry
    Bubble Gum
    Blue Moonshine
    Buddha's Sister
    Chocolope
    Durban Poison
    G-13 x Haze
    Grapefruit
    Hash Plant
    Hawaiian
    Hawaiian Skunk
    Hawaiian Snow
    Headband
    Jilly Bean
    Mango Haze
    Maui Wowie
    Northern Lights
    NYC Diesel
    OG Kush
    Purple Haze
    Skunk
    Sour Diesel
    Strawberry Cough
    Super Silver Haze
    White Widow

Anti-Fatigue

Flo
Island Sweet Skunk

Anti-Inflammatory

Champagne
Diablo OG
Durban Poison
G-13
Girl Scout Cookies
God's Gift
Great White Shark
Hawaiian Snow
Lambsbread
Mango Haze
Strawberry Diesel
White Rhino

Anxiety

Afghan Goo (aka Afghooey)
Big Bud
Blackberry Kush
Blue Moonshine
Blueberry
Bruce Banner
Bubba Kush
Bubble Gum
Champagne
Chem Dog (aka Chem Dawg)
Diablo OG

Anxiety (cont.)

Durban Poison
Flo
G-13 x Haze
Girl Scout Cookies
Grapefruit
Great White Shark
Hash Plant
Hawaiian
Hawaiian Skunk
Jack Flash
Jack Herer
Jilly Bean
Juicy Fruit
LA Confidential
Lavender Kush
Mango Haze
Master Kush
Northern Lights
OG Kush
Purple Haze
Shiva Shanti
Silver Haze
Skunk
Skywalker
Sour Diesel
Strawberry Diesel
Super Silver Haze
Sweet Tooth
Train Wreck
Vanilla Kush

Ailments & Strains

Anxiety (cont.)
    White Rhino
    White Widow

Appetite Stimulation
    Afghan Goo (aka Afghooey)
    Afghooey Train Wreck
    AK-47
    Big Bud
    Blackberry Kush
    Blue Dream
    Blue Moonshine
    Bubble Berry
    Bubble Gum
    Buddha's Sister
    Diablo OG
    Durban Poison
    Flo
    G-13 x Haze
    Grand Daddy Purple
    Grapefruit
    Great White Shark
    Hash Plant
    Hawaiian Skunk
    Island Sweet Skunk
    Jack Herer
    Jilly Bean
    LA Confidential
    Lambsbread
    Lavender Kush
    Lemon Skunk

Appetite Stim (cont.)
    Master Kush
    NYC Diesel
    OG Kush
    Purple Haze
    Shiva Shanti
    Silver Haze
    Skunk
    Strawberry Diesel
    Super Lemon Haze
    Super Silver Haze
    Sweet Tooth
    Train Wreck
    Vanilla Kush
    White Rhino
    White Widow

Arthritis Pain
    Afghan Goo (aka Afghooey)
    Blue Dream
    Bruce Banner
    Chocolope
    Diablo OG
    G-13
    G-13 x Haze
    Great White Shark
    Headband
    LA Confidential
    Lambsbread
    Maui Wowie
    Shiva Shanti

Arthritis Pain (cont.)
Space Queen
Strawberry Diesel
Train Wreck

Asthma
Maui Wowie
Master Kush

Attentive Mind Set
Jack Herer
Super Lemon Haze

Autism
Lemon Skunk

Back Pain
Northern Lights
Strawberry Cough
Super Silver Haze
Vanilla Kush

Bipolar Disorder
AK-47
Champagne
Hash Plant
Lavender Kush
Maui Wowie
Purple Haze

Body Aches
Buddha's Sister

Cachexia
White Widow

Cancer
Afghan Goo (aka Afghooey)
Blue Dream
Champagne
Chocolope
Durban Poison
G-13
God's Gift
Headband
Lambsbread
Lemon Skunk
Purple Haze
White Widow

Chemotherapy
Afghan Goo (aka Afghooey)
Blue Dream
Champagne
Chocolope
Diablo OG
Durban Poison
G-13
God's Gift
Headband
Lambsbread

Ailments & Strains

Chemotherapy (cont.)
Lemon Skunk
Purple Haze
White Widow

Chronic Pain
AK-47
Blueberry
Chernobyl
Chocolope
Diablo OG
G-13 x Haze
God's Gift
Great White Shark
Hash Plant
Jack Flash
LA Confidential
Shiva Shanti
Skywalker
Space Queen
Strawberry Diesel
White Rhino

Cramps
Afghooey Train Wreck
Great White Shark
Purple Haze
Super Silver Haze
Vanilla Kush

Creative Thoughts
Master Kush

Crohn's Disease
Great White Shark

Deep Muscle Pain
Chem Dog (aka Chem Dawg)

Depression
Afghan Goo (aka Afghooey)
Afghooey Train Wreck
AK-47
Big Bud
Blue Dream
Bubble Berry
Bubble Gum
Blue Moonshine
Bruce Banner
Buddha's Sister
Champagne
Chernobyl
Chocolope
Durban Poison
G-13 x Haze
Girl Scout Cookies
God's Gift
Grapefruit
Hash Plant
Hawaiian

Depression (cont.)

Hawaiian Skunk
Hawaiian Snow
Headband
Jilly Bean
Lambsbread
Mango Haze
Maui Wowie
Northern Lights
NYC Diesel
OG Kush
Purple Haze
Skunk
Sour Diesel
Space Queen
Strawberry Cough
Super Silver Haze
Sweet Tooth
Vanilla Kush
White Widow

Diabetic Neuropathy

Train Wreck

Eating Disorders

Banana Kush
Mango Haze

Edema

Sour Diesel

Energy

Flo
Jack Herer
Juicy Fruit
Master Kush

Epilepsy

Afghan Goo (aka Afghooey)
Blue Dream
Bubble Berry
Buddha's Sister
Champagne
G-13 x Haze
Girl Scout Cookies
Hawaiian
Headband
Island Sweet Skunk
Lemon Skunk
Shiva Shanti
Silver Haze
Skunk
Sour Diesel
White Widow

Fibromyalgia

Afghooey Train Wreck
Banana Kush
Champagne
G-13
Great White Shark
Island Sweet Skunk

Ailments & Strains

Fibromyalgia (cont.)
Jack Herer
Sour Diesel
White Widow

Focus
Super Lemon Haze
Jack Herer

Gastroenteritis
Strawberry Cough

Gastrointestinal Issues
Afghan Goo (aka Afghooey)
Blueberry
Blue Dream
Champagne
G-13 x Haze
Girl Scout Cookies
Great White Shark
LA Confidential
Lambsbread

Glaucoma
Afghan Goo (aka Afghooey)
Banana Kush
Blue Dream
Bruce Banner
Bubble Gum
Champagne

Glaucoma (cont.)
Chem Dog (aka Chem Dawg)
Chernobyl
Chocolope
Durban Poison
G-13
Girl Scout Cookies
Hawaiian
Hawaiian Snow
Headband
Island Sweet Skunk
Lambsbread
Maui Wowie
Purple Haze
Shiva Shanti
Space Queen
Strawberry Diesel
Super Silver Haze

Headaches
AK-47
Bubba Kush
Chocolope
Hash Plant
Island Sweet Skunk
OG Kush

Hepatitis C
White Widow

HIV/AIDS

Afghan Goo (aka Afghooey)
Blue Dream
Champagne
G-13
Girl Scout Cookies
Headband
Lambsbread
Purple Haze
White Widow

Hypertension

Grapefruit
Northern Lights

Inflammation

Champagne
Diablo OG
Durban Poison
G-13
Girl Scout Cookies
God's Gift
Great White Shark
Hawaiian Snow
Lambsbread
Mango Haze
Strawberry Diesel
White Rhino

Insomnia

Afghan Goo (aka Afghooey)
Banana Kush
Big Bud
Blackberry Kush
Blueberry
Bruce Banner
Bubble Berry
Bubba Kush
Blue Moonshine
Buddha's Sister
Chernobyl
Diablo OG
G-13 x Haze
God's Gift
Grapefruit
Grand Daddy Purple
Hawaiian Skunk
LA Confidential
Lavender Kush
Lemon Skunk
Northern Lights
Purple Haze
Shiva Shanti
Skunk
Skywalker
Sweet Tooth
Train Wreck
White Rhino

Inspiration
Juicy Fruit

Joint Pain
Afghooey Train Wreck
Banana Kush
Chem Dog (aka Chem Dawg)
Durban Poison
G-13
Grand Daddy Purple
Great White Shark
Hawaiian Snow
Purple Haze
Strawberry Cough
Super Silver Haze
Vanilla Kush

Lou Gehrig's Disease
Buddha's Sister
Jack Flash

Migraines
Afghan Goo (aka Afghooey)
Afghooey Train Wreck
Big Bud
Bruce Banner
Bubble Berry
Champagne
Chocolope
Durban Poison
G-13 x Haze

Migraines (cont.)
Girl Scout Cookies
Grand Daddy Purple
Hash Plant
Hawaiian Skunk
Hawaiian Snow
Headband
Jilly Bean
Lambsbread
Lavender Kush
Mango Haze
Maui Wowie
NYC Diesel
Shiva Shanti
Sour Diesel
Space Queen
Strawberry Cough
Super Silver Haze
Vanilla Kush
White Rhino

Mood Elevation
Chem Dog (aka Chem Dawg)
Flo
Hawaiian
Island Sweet Skunk
Juicy Fruit
OG Kush

Motivation
    Juicy Fruit

Movement Disorders
    Chem Dog (aka Chem Dawg)

MS
    Afghan Goo (aka Afghooey)
    Banana Kush
    Big Bud
    Blue Dream
    Bubble Gum
    Diablo OG
    Durban Poison
    G-13
    Girl Scout Cookies
    God's Gift
    Great White Shark
    Headband
    Island Sweet Skunk
    LA Confidential
    Lambsbread
    NYC Diesel
    Purple Haze
    Purple Haze
    Shiva Shanti
    Silver Haze
    Super Silver Haze
    White Rhino

Multiple Sclerosis
    Afghan Goo (aka Afghooey)
    Banana Kush
    Big Bud
    Blue Dream
    Bubble Gum
    Diablo OG
    Durban Poison
    G-13
    Girl Scout Cookies
    God's Gift
    Great White Shark
    Headband
    Island Sweet Skunk
    LA Confidential
    Lambsbread
    NYC Diesel
    Purple Haze
    Purple Haze
    Shiva Shanti
    Silver Haze
    Super Silver Haze
    White Rhino

Muscle Pain
    Afghan Goo (aka Afghooey)
    Afghooey Train Wreck
    Banana Kush
    Blueberry
    Blue Dream
    Bubble Berry

Muscle Pain (cont.)
- Chem Dog (aka Chem Dawg)
- Durban Poison
- Flo
- G-13
- Headband
- Lemon Skunk
- Purple Haze
- Shiva Shanti
- Skywalker
- Super Silver Haze
- Vanilla Kush
- White Rhino

Muscle Spasms
- Afghan Goo (aka Afghooey)
- Banana Kush
- Blackberry Kush
- Blue Dream
- Bubble Gum
- Chernobyl
- G-13
- Girl Scout Cookies
- Headband
- Jack Flash
- Purple Haze
- Shiva Shanti
- Space Queen

Muscle Tension
- Bubble Berry
- Chernobyl
- Diablo OG
- Flo
- Space Queen
- White Rhino

Nausea
- Afghan Goo (aka Afghooey)
- Afghooey Train Wreck
- AK-47
- Banana Kush
- Big Bud
- Blackberry Kush
- Blue Dream
- Blue Moonshine
- Blueberry
- Bruce Banner
- Bubble Gum
- Buddha's Sister
- Chem Dog (aka Chem Dawg)
- Chernobyl
- Chocolope
- Durban Poison
- G-13
- Girl Scout Cookies
- God's Gift
- Grand Daddy Purple
- Hash Plant

Nausea (cont.)

Hawaiian Snow
Headband
Island Sweet Skunk
Jack Flash
Jilly Bean
Lambsbread
Lavender Kush
Mango Haze
Master Kush
Northern Lights
Shiva Shanti
Skunk
Skywalker
Sour Diesel
Space Queen
Strawberry Diesel
White Rhino

Nervousness

Hash Plant
Jack Herer

Neuropathic Pain

Durban Poison
G-13

Ocular Attention

Flo
Juicy Fruit
OG Kush

Pain

Afghan Goo (aka Afghooey)
Afghooey Train Wreck
AK-47
Banana Kush
Big Bud
Blackberry Kush
Blue Dream
Blue Moonshine
Blueberry
Bruce Banner
Bubba Kush
Bubble Berry
Champagne
Chem Dog (aka Chem Dawg)
Chernobyl
Chocolope
Durban Poison
Flo
G-13
G-13 x Haze
Girl Scout Cookies
Grand Daddy Purple
Hash Plant
Hawaiian
Hawaiian Skunk
Hawaiian Snow
Headband
Island Sweet Skunk
Jack Herer
Juicy Fruit

Pain (cont.)

LA Confidential
Lambsbread
Lavender Kush
Lemon Skunk
Mango Haze
Northern Lights
Purple Haze
Shiva Shanti
Skunk
Skywalker
Space Queen
Strawberry Cough
Super Silver Haze
Sweet Tooth
White Rhino

Panic Attacks

Shiva Shanti

Parkinson's Disease

Girl Scout Cookies
God's Gift

Phantom Limb Pain

Banana Kush
Durban Poison

PMDD

Champagne
Purple Haze

PMDD (cont.)

Super Silver Haze
Vanilla Kush

PMS

Afghan Goo (aka Afghooey)
Champagne
Purple Haze
Shiva Shanti
Super Silver Haze
Vanilla Kush

Possible Ocular Relief

Grapefruit
Super Lemon Haze

PTSD

Afghan Goo (aka Afghooey)
Blue Dream
Champagne
Girl Scout Cookies
Headband
Lambsbread
Lavender Kush
Maui Wowie
Northern Lights
Purple Haze
Strawberry Cough
Sweet Tooth
White Widow

Radiculopathy
Sour Diesel

Relaxation
Flo
Grapefruit
Grand Daddy Purple
Hash Plant
Master Kush
Shiva Shanti

Restless Behaviors
Chem Dog (aka Chem Dawg)

Seizures
Champagne
Hawaiian
Lemon Skunk
White Widow

Skin Irritation
G-13

Sleep Aid (Insomnia)
Afghan Goo (aka Afghooey)
Banana Kush
Big Bud
Blueberry
Bubble Berry
Bubba Kush

Sleep Aid (cont.)
Blue Moonshine
Buddha's Sister
G-13 x Haze
Grapefruit
Grand Daddy Purple
Hawaiian Skunk
LA Confidential
Lavender Kush
Lemon Skunk
Northern Lights
Purple Haze
Shiva Shanti
Skunk
Skywalker
Train Wreck
White Rhino

Social Anxiety
Jack Herer

Soreness
Bubba Kush

Stiff Muscles
Bubba Kush

Stomach Ailments
Big Bud
Bubble Berry
NYC Diesel

Stomach Ailments (cont.)
OG Kush
White Rhino

Stomach Pain
Big Bud
Bubble Berry
NYC Diesel
OG Kush
White Rhino

Stomach Ulcers
Bubble Berry

Stress
Big Bud
Blackberry Kush
Bubble Berry
Bubble Gum
Blue Moonshine
Chernobyl
God's Gift
Hash Plant
Hawaiian Skunk
Strawberry Diesel
White Rhino

Strong Ocular Attention
Flo
Juicy Fruit
OG Kush

Tension
Bubble Berry
Flo
Skunk
Strawberry Diesel
Sweet Tooth
White Rhino

Tourette's Syndrome
Big Bud
Jack Flash
Maui Wowie

Vomiting (Nausea)
Afghan Goo (aka Afghooey)
Afghooey Train Wreck
AK-47
Banana Kush
Big Bud
Blackberry Kush
Blue Dream
Blue Moonshine
Blueberry
Bruce Banner
Bubble Gum
Buddha's Sister
Chem Dog (aka Chem Dawg)
Chernobyl
Chocolope
Durban Poison

Vomiting (cont.)

    G-13

    Girl Scout Cookies

    God's Gift

    Grand Daddy Purple

    Hash Plant

    Hawaiian Snow

    Headband

    Island Sweet Skunk

    Jack Flash

    Jilly Bean

    Lambsbread

    Lavender Kush

    Mango Haze

    Master Kush

    Northern Lights

    Shiva Shanti

    Skunk

    Skywalker

    Sour Diesel

    Space Queen

    Strawberry Diesel

    White Rhino

Cannabis Analysis Labs

Getting cannabis tested for potency, THC content, CBD content, terpenes and cannabinoid profiles can be very helpful to MMj (Medical Marijuana) patients. It can help in determining dosage.

It is also important to know if your herb has pesticides, mold or other contaminants. The only way to know for sure is to have your cannabis analyzed in a competent lab.

Several of the cannabis testing labs we listed our last edition of this book have been closed down. As of publication (August 2013), these are the testing labs we are currently aware of. There may be others that we were unable to find with our research.

NOTE: Listing a testing lab here does NOT mean we are endorsing the lab. The information below is provided for informational purposes only. We suggest that you do your own research to determine if you want to do business with a particular lab. In addition, the information provided is accurate at the time of publication, but we cannot guarantee accuracy in the future as companies move, change ownership and close their businesses.

Analytical 360, LLC
4035 Stone Way N
Seattle, WA 98103
Phone: 206-577-6998
http:// www.analytical360.com
eMail: sales@analytical360.com

Arizona Medical Marijuana Testing
Arizona
http://azmedtest.com

BudGenius.com
3905 State Street, Suite 7504
Santa Barbara, CA, 93105
Phone: 855-SAFE-BUD
http://budgenius.com
eMail: genius@budgenius.com

CannabAnalysis Laboratories
Missoula, MT
Phone: 406-531-6726
http://www.cannabanalysis.com
rose@cannabanalysis.com

CannaChemistry
Los Angeles Area
Phone: 310-221-1635
http://www.cannachemistry.com
Contact: Jeff Brown, President & CEO
eMail: jeff@cannachemistry.com

Cannatest
Seattle and Denver
Mobile Laboratory
Phone: 206-920-4427
Phone: 206-245-5800
eMail: Klaas@canna-test.com
eMail: Derek@canna-test.com

Cannabis Testing Labs

CannLabs, Inc.
3888 E Mexico Ave, Suite 238
Denver Colorado 80210
Phone: 303-309-0105
http://www.cannlabs.com/
eMail: info@cannlabs.com

CW Analytical Laboratories
Oakland, CA
Phone: 510-545-6984
www.cwanalytical.com
lab@cwanalytical.com

Green Leaf Lab (Portland, OR)
12025 NE Marx, Unit H, Building 5
Portland, Oregon 97220
Phone: 503-250-2912
http://www.greenleaflab.org
eMail: info@greenleaflab.org

Green Leaf Lab (Tacoma, WA)
1912 Center St Ste B
Tacoma WA 98409
Phone: 253-772-8771
http://www.greenleaflab.org
eMail: info@greenleaflab.org

Halent Laboratories
Northern California
Phone: 530-400-9586
http://www.halent.com
eMail: info@halent.com

Iron Laboratories

706 N. Pontiac Trail

Walled Lake, MI 48390-3408

Phone: 248-757-TEST

http://www.ironlabsllc.co

eMail: howard.lutz@ironlabsllc.co

Northwest Botanical Analysis

127 N 35th St

Seattle, WA, 98103

Phone: 206-545-7233

http://www.nwbotanicalanalysis.com

eMail: contact@nwbotanicalanalysis.com

Pure Analytics

Northern California

Phone: 888-505-7108

http://www.pureanalytics.net

eMail: info@pureanalytics.net

SC Laboratories

Monterey County California

Phone: 831-475-1844

http://www.scanalytical.com

eMail: info@sclabscom

Sequoia Analytical Labs

Sacramento, CA 95838

Phone: 916-920-4009

Phone: 916-747-4009

http://www.sequoia-labs.com/

eMail: Jeff.Hatley@sequoia-labs.com

eMail: Eric.Vierria@sequoia-labs.com

Steep Hill Labs, Inc.

473 Roland Way, Suite A

Oakland, CA 94621

Phone: 510-562-7400

Fax: 510-562-7402

http://www.steephilllab.com

info@steephilllab.com

The Werc Shop

Los Angeles Area

Phone: 310-703-9567

http://thewercshop.com

eMail: Erby@TheWercShop.com

NOTE: It is unlawful to send cannabis through the mail or via a private shipping service (such as FEDEX or UPS) to a testing lab. You will have to carry your samples in for testing, or utilize a mobile lab that comes to you.

If you have questions about testing your medical marijuana, please contact one of the labs listed here, or if you know of another analysis lab that tests cannabis, contact them.

www.ingramcontent.com/pod-product-compliance
Lightning Source LLC
Chambersburg PA
CBHW070640050426
42451CB00008B/235